The Pythagorean Theorem

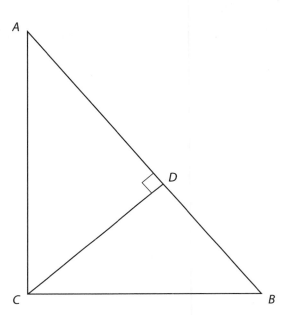

The simplest known proof of the Pythagorean theorem
(see p. 115)

The Pythagorean Theorem

❖ A 4,000-Year History

ELI MAOR

PRINCETON UNIVERSITY PRESS Princeton and Oxford

Library of Congress Cataloging-in-Publication Data
Maor, Eli.
The Pythagorean theorem : a 4,000-year history / Eli Maor.
p. cm.
Includes bibliographical references and index.
ISBN-13: 978-0-691-12526-8 (acid-free paper)
ISBN-10: 0-691-12526-0 (acid-free paper)
1. Pythagorean theorem—History. I. Title.

QA460.P8M36 2007
516.22—dc22   2006050969

British Library Cataloging-in-Publication Data is available

This book has been composed in Times

Printed on acid-free paper. ∞

pup.princeton.edu

Printed in the United States of America

1   2   3   4   5   6   7   8   9   10

To our five grandchildren:

Yehuda-Leib,

Nechama-Shira,

Yechezkel,

Zahava-Gila,

and Tzivia-Shalva.

❖ *May they enjoy a healthy, prosperous, and fulfilling life.*

Contents

Note: all photographs are courtesy of the author, except where noted.

To this day, the theorem of Pythagoras remains
the most important single theorem in the whole
of mathematics.
—Jacob Bronowski, *The Ascent of Man*, p. 160

Though its roots are in geometry, the theorem universally attributed to
Pythagoras has found its way into nearly every branch of science, pure or applied. Well over four hundred proofs of it are known, and their number is still
growing; the list includes an original proof by a future American president, another by twelve-year-old Albert Einstein, and still another by a young blind
girl. Some of these proofs are breathtaking in their simplicity, while others are
incredibly complex. The theorem itself is known by various names: *the theorem of Pythagoras*, *the hypotenuse theorem*, or simply *Euclid I 47*, so called
because it is listed as Proposition 47 in Book I of Euclid's *Elements*. Its characteristic figure (fig. P1), known in some traditions as "the windmill" and in
others as "the bride's chair," has been proposed as a cosmic identity card with
which we might introduce ourselves to extraterrestrial beings, if and when we
find them. The theorem plays a central role in numerous applications; occasionally it has been overused, even misused. And perhaps uniquely for a discipline not known for its popular appeal, it has found its way into our daily culture, appearing on postage stamps and on T-shirts, in works of art and
literature, even in the lyrics of a famous musical. By any measure, it is the
most famous theorem in all of mathematics, the one statement that every student, no matter how math-phobic, can recall from his or her high school days.

Today we think of the Pythagorean theorem as an algebraic relation,
$a^2 + b^2 = c^2$, from which the length of one side of a right triangle can be found,
given the lengths of the other two sides. But that is not how Pythagoras viewed
it; to him it was a geometric statement about areas. It was only with the rise of
modern algebra, about 1600 CE, that the theorem assumed its familiar algebraic form. It is important to bear this in mind if we are to trace the evolution
of the theorem over the 2,500 years since Pythagoras supposedly first proved
it and made it immortal. And he was not even the first to discover it: the theorem had been known to the Babylonians, and possibly to the Chinese, at least
a thousand years before him.

Many writers have commented on the beauty of the Pythagorean theorem.
Charles Lutwidge Dodgson, better known by his literary name Lewis Carroll,

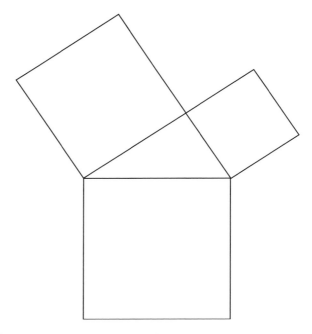

Figure P1. The Pythagorean theorem: Euclid's view

wrote in 1895: "It is as dazzlingly beautiful now as it was in the day when Pythagoras first discovered it."[1] He was certainly qualified to say this, being a talented mathematician besides gaining fame as the author of *Alice's Adventures in Wonderland* and *Through the Looking-Glass*. But who is to say what is beautiful? In 2004, the journal *Physics World* asked readers to nominate the twenty most beautiful equations in science. The top winner was Euler's formula $e^{i\pi} + 1 = 0$, followed in order by Maxwell's four electromagnetic field equations, Newton's second law of motion $F = ma$, and $a^2 + b^2 = c^2$, the Pythagorean theorem; it won only fourth place.[2]

Note that the contest was for the most beautiful *equations*, not the laws or theorems they represent. Beauty, of course, is a subjective attribute, but there is a fairly broad consensus among mathematicians as to what qualifies a theorem, or the proof thereof, to be called beautiful. A paramount criterion is symmetry. Consider, for example, the three altitudes of a triangle: they always meet at one point (as do the medians and the angle bisectors). This statement has a certain elegance to it, with its sweeping symmetry: no side or vertex takes precedence over any other; there is a complete democracy among the constituents. Or consider the theorem: If through a point P inside a circle a chord AB is drawn, the product $PA \times PB$ is constant—it has the same value for all chords through P (fig. P2). Again we have perfect democracy: every chord has the same status in relation to P as any other.

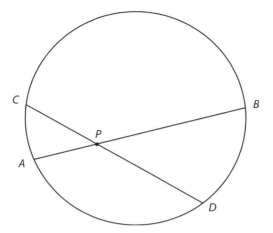

Figure P2. *PA* × *PB* = *PC* × *PD*

In this sense, the Pythagorean theorem is decidedly undemocratic. In the first place, it applies only to a very special case, that of a right triangle; and even then it singles out one side, the hypotenuse, as playing a distinctly different role from the other two sides. The word *hypotenuse* comes from the Greek words *hypo*, meaning "under," "beneath," or "down," and *teinen*, "to stretch"; this makes sense if we view the triangle with the hypotenuse at the bottom, the way it appears in Euclid's *Elements* (see again fig. P1). The Chinese call it *hsien*, a string stretched between two points (as in a lute). The Hebrew word for hypotenuse is *'yeter*, which may derive either from *mei'tar*, a string, or from *yo'ter*, "more than" (the length of each leg). But even if we look at the triangle through modern eyes, with one leg placed horizontally and the other vertically (fig. P3), the square on the hypotenuse leaps out of the figure at an odd angle. A beautiful theorem? Perhaps, but not exactly a candidate for Miss America.

If not elegance, what then is it that gives the Pythagorean theorem its universal appeal? Part of it, no doubt, has to do with the great number of proofs that have been proposed over the centuries. Elisha Scott Loomis (1852–1940), an eccentric mathematics teacher from Ohio, spent a lifetime collecting all known proofs—371 of them—and writing them up in *The Pythagorean Proposition* (1927).[3] Loomis claimed that in the Middle Ages, it was required that a student taking his Master's degree in mathematics offer a new and original proof of the Pythagorean theorem; this, he claimed, had spurred students and teachers to come up with ever new and innovative proofs. Some of these proofs are based on the similarity of triangles, others on dissection, still others on algebraic formulas, and a few make use of vectors. There are even "proofs" ("demonstrations" would be a better word) based on physical devices; in a science museum in Tel Aviv, Israel, I saw a demonstration in which colored liquid flowed freely between the squares built on the hypotenuse and on the two

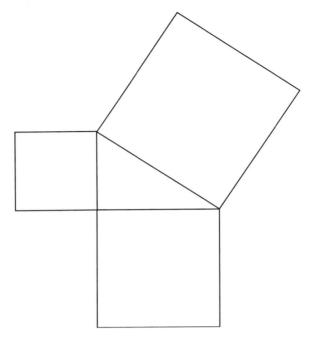

Figure P3. The Pythagorean theorem: a modern view

sides of a rotating, plexiglass-made right triangle, showing that the volume of liquid in the first square equals the combined volume in the other two.

But there is another reason for the universal appeal of the Pythagorean theorem, for it is arguably the most frequently used theorem in all of mathematics. Open any handbook of mathematical formulas; you will find the expression $x^2 + y^2$ in nearly every chapter, often tucked inside a larger expression; and it is almost always $x^2 + y^2$, not $x^3 + y^3$ or any other power of the variables. Directly or indirectly, this expression can be traced to the Pythagorean theorem. Take, for example, trigonometry, a subject notorious for its seemingly endless supply of formulas. Whether it is $\sin^2 x + \cos^2 x = 1$, or $1 + \tan^2 x = \sec^2 x$, or $1 + \cot^2 x = \csc^2 x$, these identities are the ghosts of the Pythagorean theorem—indeed, they are called the *Pythagorean identities*. The same is true in almost every branch of mathematics, from number theory and algebra to calculus and probability: in all of them, the Pythagorean theorem reigns supreme.

In this book I have traced the evolution of the Pythagorean theorem and its impact on mathematics and on our culture in general, starting with the Babylonians nearly four thousand years ago and continuing up to our own time. I have not attempted to give a comprehensive account of the hundreds of existing proofs—a nearly impossible task, and a fruitless one too, as many of these proofs are but slight variations of one another. Even Loomis's monumental compilation remains incomplete; many new proofs have been proposed since the second edition of his book appeared in 1940 (the year of his death), and new ones continue to be offered even at the time of this writing.[4]

As with my previous books, this one is aimed at the reader with an interest in the history of mathematics. Mostly, a good knowledge of high school algebra and geometry, and an occasional smattering of calculus, will be sufficient. Several subjects that require more detailed mathematical treatment have been relegated to the appendixes. Because I am making occasional reference to my earlier books, I will refer to them simply by their titles: *To Infinity and Beyond: A Cultural History of the Infinite* (1991), *e: the Story of a Number* (1994), and *Trigonometric Delights* (1998; all published by Princeton University Press, Princeton, N.J.). Two other frequently mentioned sources are Howard Eves, *An Introduction to the History of Mathematics* (Fort Worth, Texas: Saunders, 1992), and David Eugene Smith, *History of Mathematics*, vol. 1: *General Survey of the History of Elementary Mathematics*; vol. 2: *Special Topics of Elementary Mathematics* (New York, 1923–1925; rpt. New York: Dover, 1958). These will be referred to as Eves and Smith, respectively.

Many thanks go to my dear wife Dalia for encouraging me to see this work through and for her meticulous proofreading of the manuscript; to Robert Langer, for his critical review of the text and his very useful suggestions; to Vickie Kearn, my editor at Princeton University Press, for her unwavering support and encouragement to guide this book from its inception to its completion; to Debbie Tegarden, Carmina Alvarez, and Dimitri Karetnikov, and to all at the Press for their good care of the manuscript during its production phase; to Alice Calaprice, my trusted copy editor for the past fifteen years; to Joseph L. Teeters for providing me with some hard-to-find sources of useful information; to Howard Zvi Weiss for his help in translating several verses of poetry from the German; to Barbara and Jeff Niemic and to Deborah Ward for their special effort to locate and photograph the plaque in Dublin, Ireland, commemorating Sir William Rowan Hamilton's discovery of the law of quaternion multiplication; and to the staff of the Skokie Public Library in Illinois for their efforts to locate a number of obscure sources. Their help is greatly appreciated.

July 2006

Notes and Sources

1. *A New Theory of Parallels* (London, 1895).
2. *New York Times*, Ideas and Trends, October 24, 2004, p. 12.

3. (Washington, D.C.: National Council of Teachers of Mathematics, 1968.) More on this work will be found in chapter 8.

4. Several Web sites are devoted to the Pythagorean theorem and give an account of recent proofs. The Bibliography gives a partial listing of these sites.

The Pythagorean Theorem

Cambridge, England, 1993

Remember Pythagoras?
—*New York Times*, June 24, 1993

$\mathbf{M}$athematical news rarely makes the headlines, let alone front-page coverage, but June 24, 1993, was an exception. On that day, the *New York Times* ran a front-page story headed, "At Last, Shout of 'Eureka!' in Age-Old Math Mystery." Across the Atlantic the day before, a forty-year-old English mathematician announced that he had solved math's most celebrated problem, a simple-looking proposition that had kept mathematicians busy for the past 350 years.

The mathematician at the center of the excitement was Dr. Andrew Wiles, a native of Cambridge, England, and a professor at Princeton University in New Jersey. He made the sensational announcement at the end of a three-lecture series entitled "Modular Forms, Elliptic Curves, and Galois Representations." The subject was not a household term even among mathematicians, let alone laypeople. But there were rumors that the speaker would pull a surprise, and the lecture hall was packed with listeners. As the talk drew to its conclusion, the tension in the audience was palpable. Then, almost casually, Dr. Wiles ended his lecture with these words: "And by the way, this means that Fermat's Last Theorem was true. Q.E.D."[1] There was a rush to the nearest computer terminals, and those with access to e-mail services—still a novelty in 1993—flashed the news around the globe.

The circumstances behind Wiles's announcement had all the hallmarks of a human drama. Pierre de Fermat (1601–1665), a French lawyer by profession who practiced mathematics as a pastime, made a conjecture in 1637 about the possible solutions of the simple-looking equation $x^n + y^n = z^n$, where all the variables, including the exponent n, stand for positive integers. When $n = 1$, the equation is trivial: the sum of any two integers is obviously a third integer, so we have $x^1 + y^1 = z^1$. The case $n = 2$ is of greater interest. There are many triples of integers (x, y, z) for which $x^2 + y^2 = z^2$, in fact infinitely many; two examples are (3, 4, 5) and (5, 12, 13). Such triples, of course, immediately remind us of the Pythagorean theorem: they represent right triangles in which all three sides have integer lengths. So it was only natural that mathematicians tried to go to the next step—find integer solutions of the equations $x^3 + y^3 = z^3$, $x^4 + y^4 = z^4$, and so on. None were ever found.

Fermat thought he had a proof that no integer solutions of the equation $x^n + y^n = z^n$ exist for any value of n greater than 2. In the margins of his copy of the works of the third-century CE mathematician Diophantus of Alexandria, Fermat scribbled a few words that would become immortal:

> To divide a cube into two cubes, a fourth power, or in general any power whatever into two powers of the same denomination above the second is impossible. I have found an admirable proof of this, but the margin is too narrow to contain it.[2]

For the next 350 years, numerous mathematicians, laypeople, and cranks tried to reconstruct Fermat's "admirable proof." All of them failed. Two huge monetary awards, one by the French Academy of Sciences and another by its German counterpart, were offered to the first person to come up with a valid proof; both remained unclaimed.[3]

Not that the quest for a proof was entirely futile. The great Swiss mathematician Leonhard Euler (1707–1783) in 1753 proved Fermat's claim for the special case $n = 3$. Other special cases followed, and with the advent of electronic computers, all cases for n under 100,000 have been proven to be correct. But that is not the same as proving it for *all* values of n. Fermat's Last Theorem (FLT), as it became known, remained unresolved.[4]

When Wiles jumped into the fray, he already had something to start from: in 1954, a Japanese mathematician, Yutaka Taniyama (1927–1958), made a conjecture about a class of objects called *elliptic curves*. Subsequent work, particularly by Dr. Gerhard Frey of the University of the Saarland in Germany and Dr. Kenneth Ribet of the University of California at Berkeley, showed a clear connection between Taniyama's conjecture and Fermat's Last Theorem: if the former is true, then so is the latter. Wiles, after working in his attic in near seclusion for seven years, showed that the Taniyama conjecture was indeed true; and almost as an afterthought, so was FLT.

But not all was well. After submitting a 200-page-long proof to the scrutiny of mathematicians able and willing to sift through it, a tiny hole in the logic was found. Undeterred, Wiles went back into seclusion, and after another year of hard work, with the help of Cambridge lecturer Richard Taylor, he managed to fix the hole. FLT is now considered proven, finally worthy of being called a theorem.[5]

But why was this one problem singled out as the most famous unsolved problem in mathematics? For one, there was its deceiving simplicity: any high school student would be able to understand it. And the mystery of Fermat's enigmatic note only added spice to the story (most mathematicians are convinced he did not have a valid proof; the tools needed to crack the problem simply were not available in his time). But beyond these reasons, FLT leaves us with a sense that history was closing a circle. For the very same type of equation that Fermat was investigating had already been studied by the Babylonians nearly four thousand years earlier. It is here that our story really begins.

Notes and Sources

1. This is a free quotation based on the *New York Times* article of June 24, 1993, p. D22. Wiles's exact words were not reported.

2. Fermat's famous scribble, originally written in Latin, has appeared in numerous English versions. The one used here is from Eves, p. 355.

3. The French award, a gold medal and 300 francs, was offered twice, in 1815 and again in 1860. Its German counterpart was announced in 1908 and amounted to 100,000 marks—a huge sum at the time. This sum has been reduced in value by the 1929 German inflation to a paltry 7,500 marks (about $4,400 in today's value). The two prizes brought in thousands of claims, many by amateurs and cranks with little or no knowledge of mathematics.

4. The name is a misnomer in two respects: until Wiles's proof, the "theorem" was really a conjecture; and it was not Fermat's last, but rather the last of his many conjectures that mathematicians were unable to resolve.

5. Needless to say, the description of FLT given here is only the briefest of sketches. For a more detailed account, see Simon Singh's excellent book, *Fermat's Enigma: The Epic Quest to Solve the World's Greatest Mathematical Problem* (New York: Walker, 1997).

Mesopotamia, 1800 BCE

We would more properly have to call
"Babylonian" many things which the Greek
tradition had brought down to us as
"Pythagorean."
—Otto Neugebauer, quoted in Bartel van der Waerden,
Science Awakening, p. 77

The vast region stretching from the Euphrates and Tigris Rivers in the east to the mountains of Lebanon in the west is known as the Fertile Crescent. It was here, in modern Iraq, that one of the great civilizations of antiquity rose to prominence four thousand years ago: Mesopotamia. Hundreds of thousands of clay tablets, found over the past two centuries, attest to a people who flourished in commerce and architecture, kept accurate records of astronomical events, excelled in the arts and literature, and, under the rule of Hammurabi, created the first legal code in history. Only a small fraction of this vast archeological treasure trove has been studied by scholars; the great majority of tablets lie in the basements of museums around the world, awaiting their turn to be deciphered and give us a glimpse into the daily life of ancient Babylon.

Among the tablets that have received special scrutiny is one with the unassuming designation "YBC 7289," meaning that it is tablet number 7289 in the Babylonian Collection of Yale University (fig. 1.1). The tablet dates from the Old Babylonian period of the Hammurabi dynasty, roughly 1800–1600 BCE. It shows a tilted square and its two diagonals, with some marks engraved along one side and under the horizontal diagonal. The marks are in cuneiform (wedge-shaped) characters, carved with a stylus into a piece of soft clay which was then dried in the sun or baked in an oven. They turn out to be numbers, written in the peculiar Babylonian numeration system that used the base 60. In this *sexagesimal system*, numbers up to 59 were written in essentially our modern base-ten numeration system, but without a zero. Units were written as vertical Y-shaped notches, while tens were marked with similar notches written horizontally. Let us denote these symbols by | and —, respectively. The number 23, for example, would be written as — — | | |. When a number exceeded 59,

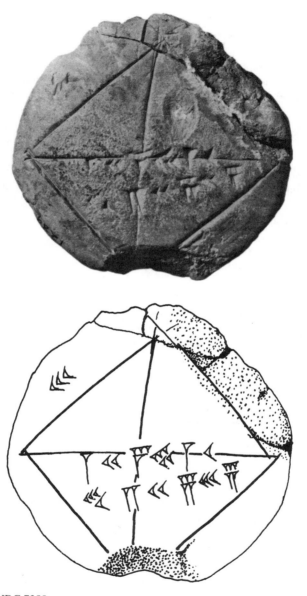

Figure 1.1. YBC 7289

it was arranged in groups of 60 in much the same way as we bunch numbers into groups of ten in our base-ten system. Thus, 2,413 in the sexagesimal system is $40 \times 60 + 13$, which was written as ————— —||| (often a group of several identical symbols was stacked, evidently to save space).

Because the Babylonians did not have a symbol for the "empty slot"—our modern zero—there is often an ambiguity as to how the numbers should be grouped. In the example just given, the numerals ————— —||| could also stand for $40 \times 60^2 + 13 \times 60 = 144{,}780$; or they could mean $40/60 + 13 = 13.166$, or any other combination of powers of 60 with the coefficients 40 and 13. Moreover, had the scribe made the space between ————— and —||| too small, the number might have erroneously been read as ———— ——|||, that is, $50 \times 60 + 3 = 3{,}003$. In such cases the correct interpretation must be deduced from the context, presenting an additional challenge to scholars trying to decipher these ancient documents.

Luckily, in the case of YBC 7289 the task was relatively easy. The number along the upper-left side is easily recognized as 30. The one immediately under the horizontal diagonal is 1;24,51,10 (we are using here the modern notation for writing Babylonian numbers, in which commas separate the sexagesimal "digits," and a semicolon separates the integral part of a number from its fractional part). Writing this number in our base-10 system, we get $1 + 24/60 + 51/60^2 + 10/60^3 = 1.414213$, which is none other than the decimal value of $\sqrt{2}$, accurate to the nearest one hundred thousandth! And when this number is multiplied by 30, we get 42.426389, which is the sexagesimal number 42;25,35—the number on the second line below the diagonal. The conclusion is inescapable: the Babylonians knew the relation between the length of the diagonal of a square and its side, $d = a\sqrt{2}$. But this in turn means that they were familiar with the Pythagorean theorem—or at the very least, with its special case for the diagonal of a square ($d^2 = a^2 + a^2 = 2a^2$)—more than a thousand years before the great sage for whom it was named.

Two things about this tablet are especially noteworthy. First, it proves that the Babylonians knew how to compute the square root of a number to a remarkable accuracy—in fact, an accuracy equal to that of a modern eight-digit calculator.[1] But even more remarkable is the probable purpose of this particular document: by all likelihood, it was intended as an example of how to find the diagonal of *any* square: simply multiply the length of the side by 1;24,51,10. Most people, when given this task, would follow the "obvious" but more tedious route: start with 30, square it, double the result, and take the square root: $d = \sqrt{30^2 + 30^2} = \sqrt{1800} = 42.4264$, rounded to four places. But suppose you had to do this over and over for squares of different sizes; you would have to repeat the process each time with a new number, a rather tedious task. The anonymous scribe who carved these numbers into a clay tablet nearly four thousand years ago showed us a simpler way: just multiply the side of the square by $\sqrt{2}$ (fig. 1.2). Some simplification!

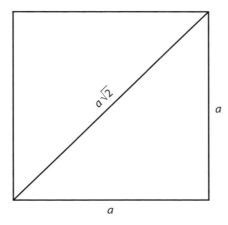

Figure 1.2. A square and its diagonal

But there remains one unanswered question: why did the scribe choose a side of 30 for his example? There are two possible explanations: either this tablet referred to some particular situation, perhaps a square field of side 30 for which it was required to find the length of the diagonal; or—and this is more plausible—he chose 30 because it is one-half of 60 and therefore lends itself to easy multiplication. In our base-ten system, multiplying a number by 5 can be quickly done by halving the number and moving the decimal point one place to the right. For example, $2.86 \times 5 = (2.86/2) \times 10 = 1.43 \times 10 = 14.3$ (more generally, $a \times 5 = \frac{a}{2} \times 10$). Similarly, in the sexagesimal system multiplying a number by 30 can be done by halving the number and moving the "sexagesimal point" one place to the right ($a \times 30 = \frac{a}{2} \times 60$).

Let us see how this works in the case of YBC 7289. We recall that 1;24,51,10 is short for $1 + 24/60 + 51/60^2 + 10/60^3$. Dividing this by 2, we get $\frac{1}{2} + \frac{12}{60} + \frac{25\frac{1}{2}}{60^2} + \frac{5}{60^3}$, which we must rewrite so that each coefficient of a power of 60 is an integer. To do so, we replace the 1/2 in the first and third terms by by 30/60, getting $\frac{30}{60} + \frac{12}{60} + \frac{25+\frac{30}{60}}{60^2} + \frac{5}{60^3} = \frac{42}{60} + \frac{25}{60^2} + \frac{35}{60^3} = 0;42,25,35$. Finally, moving the sexagesimal point one place to the right gives us 42;25,35, the length of the diagonal. It thus seems that our scribe chose 30 simply for pragmatic reasons: it made his calculations that much easier.

❖   ❖   ❖

If YBC 7289 is a remarkable example of the Babylonians' mastery of elementary geometry, another clay tablet from the same period goes even further: it shows that they were familiar with algebraic procedures as well.[2] Known as

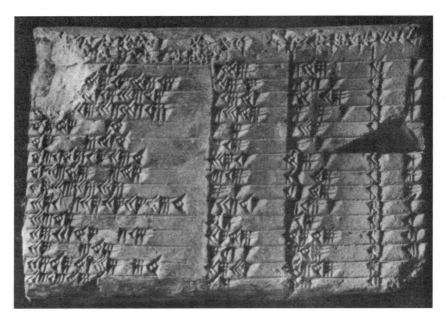

Figure 1.3. Plimpton 322

Plimpton 322 (so named because it is number 322 in the G. A. Plimpton Collection at Columbia University; see fig. 1.3), it is a table of four columns, which might at first glance appear to be a record of some commercial transaction. A close scrutiny, however, has disclosed something entirely different: the tablet is a list of *Pythagorean triples*, positive integers (a, b, c) such that $a^2 + b^2 = c^2$. Examples of such triples are (3, 4, 5), (5, 12, 13), and (8, 15, 17). Because of the Pythagorean theorem,[3] every such triple represents a right triangle with sides of integer length.

Unfortunately, the left edge of the tablet is partially missing, but traces of modern glue found on the edges prove that the missing part broke off after the tablet was discovered, raising the hope that one day it may show up on the antiquities market. Thanks to meticulous scholarly research, the missing part has been partially reconstructed, and we can now read the tablet with relative ease. Table 1.1 reproduces the text in modern notation. There are four columns, of which the rightmost, headed by the words "its name" in the original text, merely gives the sequential number of the lines from 1 to 15. The second and third columns (counting from right to left) are headed "solving number of the diagonal" and "solving number of the width," respectively; that is, they give the length of the diagonal and of the short side of a rectangle, or equivalently, the length of the hypotenuse and the short leg of a right triangle. We will label these columns with the letters c and b, respectively. As

TABLE 1.1
Plimpton 322

$(c/a)^2$	b	c	
[1,59,0,]15	1,59	2,49	1
[1,56,56,]58,14,50,6,15	56,7	3,12,1	2
[1,55,7,]41,15,33,45	1,16,41	1,50,49	3
[1,]5[3,1]0,29,32,52,16	3,31,49	5,9,1	4
[1,]48,54,1,40	1,5	1,37	5
[1,]47,6,41,40	5,19	8,1	6
[1,]43,11,56,28,26,40	38,11	59,1	7
[1,]41,33,59,3,45	13,19	20,49	8
[1,]38,33,36,36	9,1	12,49	9
1,35,10,2,28,27,24,26,40	1,22,41	2,16,1	10
1,33,45	45	1,15	11
1,29,21,54,2,15	27,59	48,49	12
[1,]27,0,3,45	7,12,1	4,49	13
1,25,48,51,35,6,40	29,31	53,49	14
[1,]23,13,46,40	56	53	15

Note: The numbers in brackets are reconstructed.

an example, the first line shows the entries $b = 1,59$ and $c = 2,49$, which represent the numbers $1 \times 60 + 59 = 119$ and $2 \times 60 + 49 = 169$. A quick calculation gives us the other side as $a = \sqrt{169^2 - 119^2} = \sqrt{14400} = 120$; hence (119, 120, 169) is a Pythagorean triple. Again, in the third line we read $b = 1,16,41 = 1 \times 60^2 + 16 \times 60 + 41 = 4601$, and $c = 1,50,49 = 1 \times 60^2 + 50 \times 60 + 49 = 6649$; therefore, $a = \sqrt{6649^2 - 4601^2} = \sqrt{23\,040\,000} = 4800$, giving us the triple (4601, 4800, 6649).

The table contains some obvious errors. In line 9 we find $b = 9,1 = 9 \times 60 + 1 = 541$ and $c = 12, 49 = 12 \times 60 + 49 = 769$, and these do not form a Pythagorean triple (the third number a not being an integer). But if we replace the 9,1 by $8,1 = 481$, we do indeed get an integer value for a: $a = \sqrt{769^2 - 481^2} = \sqrt{360\,000} = 600$, resulting in the triple (481, 600, 769). It seems that this error was simply a "typo"; the scribe may have been momentarily distracted and carved nine marks into the soft clay instead of eight; and once the tablet dried in the sun, his oversight became part of recorded history.

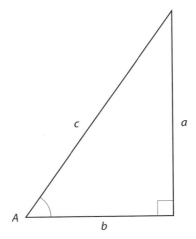

Figure 1.4. The cosecant of an angle: csc $A = c/a$

Again, in line 13 we have $b = 7,12,1 = 7 \times 60^2 + 12 \times 60 + 1 = 25\,921$ and $c = 4,49 = 4 \times 60 + 49 = 289$, and these do not form a Pythagorean triple; but we may notice that 25 921 is the square of 161, and the numbers 161 and 289 do form the triple (161, 240, 289). It seems the scribe simply forgot to take the square root of 25 921. And in row 15 we find $c = 53$, whereas the correct entry should be twice that number, that is, $106 = 1,46$, producing the triple (56, 90, 106).[4] These errors leave one with a sense that human nature has not changed over the past four thousand years; our anonymous scribe was no more guilty of negligence than a student begging his or her professor to ignore "just a little stupid mistake" on the exam.[5]

The leftmost column is the most intriguing of all. Its heading again mentions the word "diagonal," but the exact meaning of the remaining text is not entirely clear. However, when one examines its entries a startling fact comes to light: this column gives the square of the ratio c/a, that is, the value of $\csc^2 A$, where A is the angle opposite side a and csc is the cosecant function studied in trigonometry (fig. 1.4). Let us verify this for line 1. We have $b = 1,59 = 119$ and $c = 2,49 = 169$, from which we find $a = 120$. Hence $(c/a)^2 = (169/120)^2 = 1.983$, rounded to three places. And this indeed is the corresponding entry in column 4: $1;59,0,15 = 1 + 59/60 + 0/60^2 + 15/60^3 = 1.983$. (We should note again that the Babylonians did not use a symbol for the "empty slot" and therefore a number could be interpreted in many different ways; the correct interpretation must be deduced from the context. In the example just cited, we assume that the leading 1 stands for units rather than sixties.) The reader may check other entries in this column and confirm that they are equal to $(c/a)^2$.

Several questions immediately arise: Is the order of entries in the table random, or does it follow some hidden pattern? How did the Babylonians find those particular numbers that form Pythagorean triples? And why were they interested in these numbers—and in particular, in the ratio $(c/a)^2$—in the first place? The first question is relatively easy to answer: if we compare the values of $(c/a)^2$ line by line, we discover that they decrease steadily from 1.983 to 1.387, so it seems likely that the order of entries was determined by this sequence. Moreover, if we compute the square root of each entry in column 4—that is, the ratio $c/a = \csc A$—and then find the corresponding angle A, we discover that A *increases* steadily from just above 45° to 58°. It therefore seems that the author of this text was not only interested in finding Pythagorean triples, but also in determining the ratio c/a of the corresponding right triangles. This hypothesis may one day be confirmed if the missing part of the tablet shows up, as it may well contain the missing columns for a and c/a. If so, Plimpton 322 will go down as history's first trigonometric table.

As to how the Babylonian mathematicians found these triples—including such enormously large ones as (4601, 4800, 6649)—there is only one plausible explanation: they must have known an algorithm which, 1,500 years later, would be formalized in Euclid's *Elements*: Let u and v be any two positive integers, with $u > v$; then the three numbers

$$a = 2uv, \quad b = u^2 - v^2, \quad c = u^2 + v^2 \tag{1}$$

form a Pythagorean triple. (If in addition we require that u and v are of opposite parity—one even and the other odd—and that they do not have any common factor other than 1, then (a, b, c) is a *primitive* Pythagorean triple, that is, a, b, and c have no common factor other than 1.) It is easy to confirm that the numbers a, b, and c as given by equations (1) satisfy the equation $a^2 + b^2 = c^2$:

$$\begin{aligned}
a^2 + b^2 &= (2uv)^2 + (u^2 - v^2)^2 \\
&= 4u^2v^2 + u^4 - 2u^2v^2 + v^4 \\
&= u^4 + 2u^2v^2 + v^4 \\
&= (u^2 + v^2)^2 = c^2.
\end{aligned}$$

The converse of this statement—that *every* Pythagorean triple can be found in this way—is a bit harder to prove (see Appendix B).

Plimpton 322 thus shows that the Babylonians were not only familiar with the Pythagorean theorem, but that they knew the rudiments of number theory and had the computational skills to put the theory into practice—quite remarkable for a civilization that lived a thousand years before the Greeks produced their first great mathematician.

Notes and Sources

1. For a discussion of how the Babylonians approximated the value of $\sqrt{2}$, see Appendix A.

2. The text that follows is adapted from *Trigonometric Delights* and is based on

Otto Neugebauer, *The Exact Sciences in Antiquity* (1957; rpt. New York: Dover, 1969), chap. 2. See also Eves, pp. 44–47.

3. More precisely, its *converse*: if the sides of a triangle satisfy the equation $a^2 + b^2 = c^2$, the triangle is a right triangle.

4. This, however, is not a *primitive triple*, since its members have the common factor 2; it can be reduced to the simpler triple (28, 45, 53). The two triples represent similar triangles.

5. A fourth error occurs in line 2, where the entry 3,12,1 should be 1,20,25, producing the triple (3367, 3456, 4825). This error remains unexplained.

Sidebar 1

Did the Egyptians Know It?

The Egyptians must have used this formula
$[a^2 + b^2 = c^2]$ or they couldn't have built their
pyramids, but they have never expressed it as a
useful theory.
—Joy Hakim, *The Story of Science*, p. 78

Five hundred miles to the southwest of Mesopotamia, along the Nile Valley, thrived a second great ancient civilization, Egypt. The two nations coexisted in relative peace for over three millennia, from about 3500 BCE to the time of the Greeks. Both developed advanced writing skills, were keen observers of the sky, and kept meticulous records of their military victories, commercial transactions, and cultural heritage. But whereas the Babylonians recorded all this on clay tablets—a virtually indestructible writing material—the Egyptians used papyrus, a highly fragile medium. Were it not for the dry desert climate, their writings would have long been disintegrated. Even so, our knowledge of ancient Egypt is less extensive than that of its Mesopotamian contemporary; what we do know comes mainly from artifacts found in the burial sites of the ruling Egyptian dynasties, from a handful of surviving papyrus scrolls, and from hieroglyphic inscriptions on their temples and monuments.

Most famous of all Egyptian shrines are the pyramids, built over a period of 1,500 years to glorify the pharaoh rulers during their lives, and even more so after their deaths. A huge body of literature has been written on the pyramids; regrettably, much of this literature is more fiction than fact. The pyramids have attracted a cult of worshipers who found in these monuments hidden connections to just about everything in the universe, from the numerical values of π and the Golden Ratio to the alignment of planets and stars. To quote the eminent Egyptologist Richard J. Gillings: "Authors, novelists, journalists, and writers of fiction found during the nineteenth century a new topic [the pyramids], a new idea to develop, and the less that was known and clearly understood about the subject, the more freely could they give rein to their imagination."[1]

Certainly, building such a huge monument as the Great Pyramid of Cheops—756 feet on each side and soaring to a height of 481 feet—required a good deal of mathematical knowledge, and surely that knowledge must have included the Pythagorean theorem. But did it? Our main source of information on ancient Egyptian mathematics comes from the Rhind Papyrus, a collection of eighty-four problems dealing with arithmetic, geometry, and rudimentary algebra. Discovered in 1858 by the Scottish Egyptologist A. Henry Rhind, the papyrus is 18 feet long and 13 inches wide. It survived in remarkably good condition and is the oldest mathematics textbook to reach us nearly intact (it is now in the British Museum in London).[2] The papyrus was written about 1650 BCE by a scribe named A'h-mose, commonly known in the West as Ahmes. But it was not his own work; as A'h-mose tells us, he merely copied it from an older document dated to about 1800 BCE. Each of the eighty-four problems is followed by a detailed step-by-step solution; some problems are accompanied by drawings. Most likely the work was a training manual for use in a school of scribes, for it was the sect of royal scribes to whom all literary tasks were assigned—reading, writing, and arithmetic, our modern "Three R's."

Of the eighty-four problems in the Rhind Papyrus, twenty are geometric in nature, dealing with such questions as finding the volume of a cylindrical granary or the area of a field of given dimensions (this latter problem was of paramount importance to the Egyptians, whose livelihood depended on the annual inundation of the Nile). Five of these problems specifically concern the pyramids; yet not once is there any reference in them to the Pythagorean theorem, either directly or by implication. One concept that does appear repeatedly is the *slope* of the side of a pyramid, a question of considerable significance to the builders, who had to ensure that all four faces maintained an equal and uniform slope.[3] But the Pythagorean theorem? Not once.

Of course, the absence of evidence is not evidence of absence, as archeologists like to point out. Still, in all likelihood the Rhind Papyrus represented a summary of the kind of mathematics a learned person—a scribe, an architect, or a tax collector—might encounter in his career, and the absence of any reference to the Pythagorean theorem strongly suggests that the Egyptians did not know it.[4] It is often said that they used a rope with knots tied at equal intervals to measure distances; the 3-4-5 knotted rope, so the logic goes, must have led the Egyptians to discover that a 3-4-5 triangle is a right triangle and thus, presumably, to the the fact that $3^2 + 4^2 = 5^2$. But there is no evidence whatsoever to support this hypothesis. It is even less plausible that they used the 3-4-5 rope to construct a right angle, as some authors have stated; it would have been so much easier to use a plumb line for that purpose. The case is best summarized by quoting three eminent scholars of ancient mathematics:

> In 90% of all the books [on the history of mathematics], one finds the statement the Egyptians knew the right triangle of sides 3, 4 and 5, and that they used it for laying out right angles. How much value has this statement? None!
> —Bartel Leendert van der Waerden.[5]

> There is no indication that the Egyptians had any notion even of the Pythagorean Theorem, despite some unfounded stories about "harpedonaptai" [rope stretchers], who supposedly constructed right triangles with the aid of a string with 3 + 4 + 5 = 12 knots.
> —Dirk Jan Struik.[6]

> There seems to be no evidence that they knew that the triangle (3, 4, 5) is right-angled; indeed, according to the latest authority (T. Eric Peet, *The Rhind Mathematical Papyrus*, 1923), nothing in Egyptian mathematics suggests that the Egyptians were acquainted with this or any special cases of the Pythagorean Theorem.
> —Sir Thomas Little Heath.[7]

Of course, archeologists may some day unearth a document showing a square with the lengths of its side and diagonal inscribed next to them, as in YBC 7289. But until that happens, we cannot conclude that the Egyptians knew of the relation between the sides and the hypotenuse of a right triangle.

Notes and Sources

1. *Mathematics in the Time of the Pharaohs* (1972; rpt. New York: Dover, 1982), p. 237.

2. See Arnold Buffum Chace, *The Rhind Mathematical Papyrus: Free Translation and Commentary with Selected Photographs, Transcriptions, Transliterations and Literal Translations* (Reston, Va.: National Council of Teachers of Mathematics, 1979).

3. On this subject, see *Trigonometric Delights*, pp. 6–9.

4. According to Smith (vol. 2, p. 288), a papyrus of the Twelfth Dynasty (ca. 2000 BCE), discovered at Kahun, refers to four Pythagorean triples, one of which is $1^2 + (3/4)^2 = (1\frac{1}{4})^2$ (which is equivalent to the triple (3, 4, 5) when cleared of fractions). Whether these triples refer to the sides of right triangles is not known.

5. *Science Awakening*, trans. Arnold Dresden (New York: John Wiley, 1963), p. 6. Van der Waerden goes on to give the reasons for making this statement, adding that "repeated copying [of the assumption that the Egyptians used

the 3-4-5 sided triangle to lay out right angles] made it a 'universally known fact.'"

6. *Concise History of Mathematics* (New York: Dover, 1967), p. 24. Struik (1894–2000) was a Dutch-born scholar who taught at the Massachusetts Institute of Technology from 1926 to 1960. In his obituary, Evelyn Simha, director of the Dibner Institute for the History of Science and Technology at MIT, described Struik as "the instructor responsible for half the world's basic knowledge of the history of mathematics" (*New York Times*, October 26, 2000, p. A29). Active almost to the end, he died at the age of 106.

7. *The Thirteen Books of Euclid's Elements*, vol. 1 (London: Cambridge University Press, 1962), p. 352.

Pythagoras

> Number rules the universe.
> —motto of the Pythagoreans

You will find his picture in every book on the history of mathematics, an old saintly figure with a long beard and a wise expression in his eyes (fig. 2.1). But who was this revered person? The truth is, we don't know. Pythagoras is one of the most mysterious figures in history; the little we do know about him may be more fiction than fact, written by historians who lived hundreds of years later. So everything you read about him—and most certainly the image on that bearded portrait—must be taken with a grain of salt.[1]

According to tradition, Pythagoras was born around 570 BCE on the island of Samos in the Aegean Sea, just off the coast of Asia Minor (modern Turkey). Not far to the east, in the coastal town of Miletus, lived the famous philosopher Thales, the first of the long line of Greek scholars who would shape the intellectual world for the next thousand years. So we may assume—though we cannot be certain—that young Pythagoras studied under the great master, who kindled in him a passion for mathematics and philosophy. Pythagoras then traveled to the major centers of civilization of the ancient world, among them Egypt and Persia, absorbing as much as he could of their literature, religion, philosophy, and mathematics. What he learned during his sojourns left a deep impression on the young scholar.

When Pythagoras returned to his native island, he found it under the harsh reign of its ruler, Polycrates, so around 530 BCE he left Samos and settled in the Greek outpost of Croton, on the southeast coast of modern Italy. There he founded a school that was to exercise an enormous influence on subsequent generations of scholars. Under the guidance of their master, the Pythagoreans studied every intellectual discipline then in existence, particularly philosophy, mathematics, and astronomy. But theirs was more than just a school: they formed a sect, a brotherhood bound by a pledge of allegiance that was strictly enforced upon its members. The Pythagoreans vowed to keep all their discussions secret, perhaps to avoid being scorned by their enemies, of which they had many. Or perhaps this pact made it easier for them to stay aloof of the

Figure 2.1. Pythagoras of Samos

daily toil that was the lot of the vast majority of their countrymen. Whatever the reason, their secrecy had unfortunate historical consequences: it prevented any of their discussions from reaching a wider audience, at least initially. What we do know about the Pythagoreans comes almost entirely from later generations of writers, who often tried to outdo one another in glorifying the great master.[2]

Added to the paucity of original sources was the fact that the Pythagoreans followed the Oriental tradition of oral transmission of knowledge. Writing material was scarce; Egyptian papyrus was introduced to Greece around 650 BCE, but it was still in short supply in Pythagoras's time. Parchment was not yet known, and clay tablets were hardly a suitable medium on which to write long philosophical discourses. As a consequence, knowledge was passed from one generation to the next mainly by word of mouth, leaving few if any written records. Moreover, out of respect for their leader, many of the discoveries made by the Pythagoreans were attributed to Pythagoras himself, so their true discoverer may never be known.

Pythagoras's first major scientific discovery was in an unlikely subject: acoustics. As the story goes, while walking down a street one day he heard sonorous sounds coming from a blacksmith's shop. Stopping to investigate, he found out that the sound originated from the vibrations of metal sheets hit by the blacksmith's hammer: the larger the sheet, the lower the pitch of the sound it produced. Pythagoras then experimented with bells and water-filled glasses and found the same general relationship: the more massive an object, the lower the pitch of its sound.

Pythagoras, however, was not content with a mere qualitative relationship. He went on to investigate the vibrations of strings (fig. 2.2) and discovered that the pitch of their sound is inversely proportional to their length. Imagine two strings of identical material and thickness and held under the same tension,

Figure 2.2. Pythagoras discovering the laws of harmony

one string twice as long as the other. Then the shorter string will vibrate at *twice* the frequency of the long string; in musical terms, the two strings are an *octave* apart. Similarly, a length ratio of 3:2 corresponds to an interval of a *fifth*, a ratio of 4:3 to a *fourth*, and so on (the names "octave," "fifth," and "fourth" come from the position of these intervals in the musical scale). This was an important discovery, the first time a natural phenomenon was described in terms of a precise quantitative expression. In a sense, it marked the beginning of mathematical physics.

From there it was but one step to Pythagoras's next discovery. If the two strings are allowed to vibrate simultaneously, they produce a musical *chord*. To his great delight, Pythagoras found that simple ratios of string lengths produce pleasant chords, or consonants; chief among them are the intervals just mentioned—the octave, fifth, and fourth, which Pythagoras called the "perfect intervals." More complicated ratios correspond to less pleasing chords; for example, the ratio 9:8 produces a *second*, a distinctly dissonant chord. Pythagoras therefore concluded that numerical ratios rule the laws of musical harmony—and by extension, the entire universe. It was to become an idée fixe with the Pythagoreans, the cornerstone of their world picture.

To understand this giant leap of faith, we must remember that in ancient Greece, music ranked equal in importance to arithmetic (specifically, number theory), geometry, and astronomy; indeed, these four subjects would become the *quadrivium*, the core curriculum that every educated person was expected to master. And because music was put on an equal footing with mathematics, it is not surprising that each exerted a profound influence on the other (just think of the frequent occurrence of the word "harmonic" in mathematics—harmonic mean, harmonic series, and harmonic function, to name but a few).

So, the Pythagoreans reasoned, if numbers govern the laws of musical harmony, they must also govern everything else in the universe.

The Pythagorean fixation with numbers had several consequences, some beneficial to the advancement of science, others detrimental. On the one hand, it spurred them to study the mathematical properties of numbers (here meaning positive integers); and by so doing they planted the seeds that would grow into modern number theory. But like any obsession when pursued fanatically, it led to aberrations. Their uncompromising belief in the supremacy of numbers—whether in musical harmony or in the physical universe—blinded the Pythagoreans to other ideas that might have better explained the workings of nature. By subjugating the laws of nature to the Greek ideals of beauty, symmetry, and harmony, the Pythagoreans hindered the progress of the physical sciences for a good two thousand years.

Nowhere is this more clear than in Greek cosmology. The Pythagoreans thought of the Earth as a motionless globe suspended at the center of the universe, while the fixed stars, embedded like tiny jewels in the celestial dome, moved around it in perfect circles once every twenty-four hours. The Sun, Moon, and the five then-known planets revolved around the Earth in circular orbits of their own, superimposed on the motion of the celestial dome. When these circular orbits did not quite fit the observational data, they were replaced in the third century BCE by *epicycles*, small circles whose centers moved along the main circular orbit. In time the number of epicycles grew, with more and more added to the existing ones, until the system became so cumbersome as to be practically useless. Nevertheless, the Greek world picture was to dominate astronomy for over two thousand years. The possibility that the heavenly bodies might follow any orbit other than a circle was unacceptable to the Greeks: it *had* to be a circle, the most perfect of all shapes. Even when Nicolaus Copernicus, in his great work *De Revolutionibus* (On the revolutions of the heavenly bodies, 1543) dethroned the Earth from its lofty position at the center of the universe and replaced it by the Sun, he still clung to the good old circular orbits. It was not until 1609 that Johannes Kepler put the circular orbits to rest and replaced them with ellipses (which Newton later expanded to include parabolas and hyperbolas).

❖   ❖   ❖

But let us return to mathematics. The Pythagorean study of numbers began as a spiritual quest, something akin to the modern Kaballah. To them, every number carried a symbolic identity. One was the generator of all numbers, because every number can be created from it by repeated addition; consequently, it had a special status and was not considered a number proper. Two and three stood for the female and male characters, respectively, and five, for their union. Five was also the number of *regular polyhedra*, solids whose faces are identical regular polygons (fig. 2.3): the tetrahedron, with four equilateral

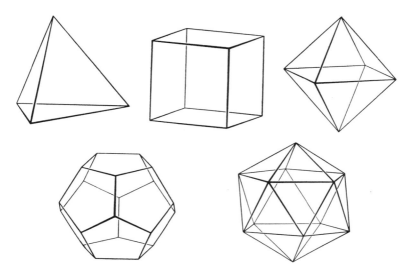

Figure 2.3. The five regular or Platonic solids

triangles, the cube (six squares), the octahedron (eight triangles), the dodeca-hedron (twelve pentagons), and the icosahedron (twenty triangles). These five solids, according to the Pythagoreans, represented the five elements of which the universe was thought to be made: fire, earth, air, water, and the heavenly dome surrounding them all. Consequently, five acquired something of a sa-cred status, and in the form of a pentagram it became the Pythagorean em-blem (fig. 2.4).

Even more sacred than five was six, the first *perfect number*. A number is perfect if it is the sum of its proper divisors (including 1 but excluding the

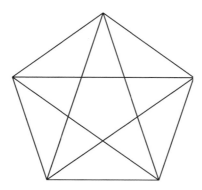

Figure 2.4. The pentagram: the Pythagorean emblem

number itself). The proper divisors of 6 are 1, 2, and 3, and since $1 + 2 + 3 = 6$, six is the first perfect number. The next three are 28 $(= 1 + 2 + 4 + 7 + 14)$, 496 $(= 1 + 2 + 4 + 8 + 16 + 31 + 62 + 124 + 248)$, and 8,128 $(= 1 + 2 + 4 + 8 + 16 + 32 + 64 + 127 + 254 + 508 + 1,016 + 2,032 + 4,064)$. These four were the only perfect numbers known to the Greeks; the fifth, 33,550,336, was only found in 1456. At the time of this writing, 43 perfect numbers are known, all of them even; it is not known whether any odd perfect numbers exist, nor whether the number of perfect numbers is finite or infinite.[3]

❖   ❖   ❖

Of particular interest to the Pythagoreans were *figurative numbers*, numbers that can be represented as dots arranged in a regular pattern. Consider the sum of the first five integers, $1 + 2 + 3 + 4 + 5$. We can represent this sum as dots arranged in a staircase:

$$1 + 2 + 3 + 4 + 5$$

To find the total number of these dots, let us fill in the missing spaces to get a rectangle:

$$1 + 2 + 3 + 4 + 5 = (5 \times 6)/2 = 15$$

This rectangle has $5 \times 6 = 30$ dots; and because this is twice the number of black dots, the required sum is one-half of 30, or 15. Generalizing to the sum of the first *n* integers, we get the formula $1 + 2 + 3 + \ldots + n = \frac{n(n+1)}{2}$.[4]

An even more interesting pattern involves the sum of the first *n odd* integers, $1 + 3 + 5 + \ldots + (2n - 1)$. We notice that $1 = 1^2$, $1 + 3 = 4 = 2^2$, $1 + 3 + 5 = 9 = 3^2$, and so on. No matter how many odd integers we add, their sum is always a *perfect square*: $1 + 2 + 3 + \ldots + (2n - 1) = n^2$.

The dot pattern below makes this clear:

$$1 = 1^2 = 1 \qquad 1 + 3 = 2^2 = 4 \qquad 1 + 3 + 5 = 3^2 = 9$$

Such explorations led the Pythagoreans to develop a kind of primitive algebra, based on the interrelation of various figures. For example, the familiar formula $(a + b)^2 = a^2 + 2ab + b^2$ can be proved geometrically by considering a square of side $(a + b)$, as shown in figure 2.5. We can dissect this square into two smaller squares of areas a^2 and b^2 and two rectangles of areas $a \times b$ and $b \times a$. But the two rectangles are congruent, so their areas are equal. Thus the total area of the dissected parts is $a^2 + 2ab + b^2$, and this is equal to the area of the original square, $(a + b)^2$. Other formulas, such as $(a - b)^2 = a^2 - 2ab + b^2$ or $(a + b)(a - b) = a^2 - b^2$, can be proved in a similar manner. This sort of geometric algebra was a precursor to our modern symbolic algebra.[5]

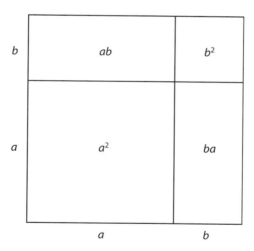

Figure 2.5. Geometric proof of $(a + b)^2 = a^2 + 2ab + b^2$

❖　❖　❖

And then there were the Pythagorean triples. As mentioned in chapter 1, these are positive integers (a, b, c) such that $a^2 + b^2 = c^2$. As the professed discoverers of the famous theorem named after their master, the Pythagoreans were naturally eager to discover right triangles with all three sides of integer length, but they soon realized that this was easier said than done: one can arbitrarily choose *two* sides with integer length, but the third side would most likely not be an integer. On rare occasions, however, such *Pythagorean triangles* were indeed found, to the sect's great delight. Legend has it that they celebrated these occasions with a feast at which one hundred oxen were sacrificed.[6]

As to how the Pythagoreans found these triangles, we have no direct evidence. According to one theory, they used the formula

$$n^2 + \left(\frac{n^2-1}{2}\right)^2 = \left(\frac{n^2+1}{2}\right)^2, \tag{1}$$

indicating that the numbers n, $\frac{n^2-1}{2}$, and $\frac{n^2+1}{2}$ form a Pythagorean triple (indeed, a primitive triple; see page 11) for every odd value of n. For example, for $n = 3$ we get the triple (3, 4, 5); for $n = 5$, the triple (5, 12, 13), and so on.

It is easy, of course, to prove equation (1) using modern algebra, but that is not how the Greeks did it. In all likelihood they used a geometric proof based on the following observation. Imagine an array of m^2 dots arranged in m rows and m columns. On the outside of this array (a square) we can place $2m + 1$ additional dots (m dots along two adjacent sides and one dot next to the corner), resulting in an extended square of $(m + 1)^2$ dots. For example, if $m = 4$, then $2m + 1 = 9$, so from an array of $4^2 = 16$ dots we get a new array of $5^2 = 16 + 9 = 25$ dots:

```
  o    o    o    o    o
  o    •    •    •    •
  o    •    •    •    •
  o    •    •    •    •
  o    •    •    •    •
```

We thus arrive at the formula

$$m^2 + (2m + 1) = (m + 1)^2, \tag{2}$$

which we recognize as a familiar identity from elementary algebra. Now suppose $2m + 1$ is a *perfect square*, say n^2. Then equation (2) becomes $m^2 + n^2 = (m + 1)^2$, producing the Pythagorean triple $(m, n, m + 1)$. Solving the equation $2m + 1 = n^2$ for m, we get $m = \frac{n^2-1}{2}$, so equation (2) becomes

$$\left(\frac{n^2-1}{2}\right)^2 + n^2 = \left(\frac{n^2+1}{2}\right)^2,$$

which is equation (1).[7]

❖   ❖   ❖

But important as these discoveries were, they pale in comparison with two events that profoundly affected the future course of mathematics: Pythagoras's proof of the famous theorem bearing his name, and the discovery of a new

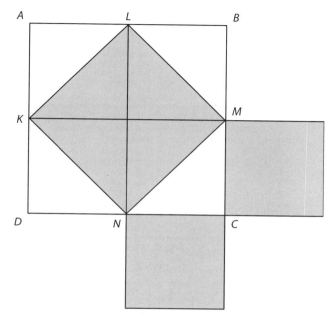

Figure 2.6. A special case of the Pythagorean theorem: the 45-45-90–degree triangle

kind of number that cannot be written as a ratio of two integers: an *irrational number*. Neither the proof nor details of the discovery have survived, so all we can do is rely on writings by later authors and add our own speculations.[8]

When Euclid wrote his *Elements* around 300 BCE, he gave two proofs of the Pythagorean theorem: one, Proposition 47 of Book I, relies entirely on area relations and is quite sophisticated; the other, Proposition 31 of Book VI, is based on the concept of proportion and is much simpler (we will discuss both proofs in chapter 3). Were these Pythagoras's own proofs? Most certainly, Pythagoras could not have used I 47; geometry was simply not advanced enough in his time. He may have used VI 31, but if so, his proof was deficient, because the complete theory of proportions was only developed by Eudoxus, who lived almost two centuries after Pythagoras. On the other hand, there is some reason to believe that Pythagoras had first proved the special case of a right isosceles triangle, that is, a 45-45-90–degree triangle. This proof was already known to the Hindus, and Pythagoras may have heard of it during his travels around the Mediterranean. The proof is fairly simple: in square *ABCD* (fig. 2.6), join the midpoints of adjacent as well as opposite sides. The inner square *KLMN* is dissected into four congruent 45-45-90–degree triangles. Any two of these have a combined area equal to that of a square built on the side *MC* or *NC* of triangle *MCN*, so the total area of square *KLMN* is equal to the combined area of the squares built on the two sides.

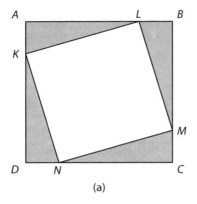
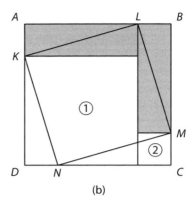

Figure 2.7. The "Chinese proof"

Did Pythagoas also prove the general case? We have no direct evidence, but it is hard to believe that, having demonstrated the special case, he would not have proceeded to establish the general theorem as well. Tradition ascribes to him a demonstration that was already known to the Chinese. By tilting the inner square in figure 2.6 from its 45-degree position, we get the configuration shown in figure 2.7a. Square *ABCD* is now dissected into an inner square *KLMN* and four congruent right triangles (shaded in the figure). By reassembling these right triangles as in figure 2.7b, we see that the remaining (unshaded) area is the sum of the areas of squares 1 and 2, that is, the squares built on the sides of each of the right triangles.

We will meet the "Chinese proof" again in chapter 5. Whether this was the demonstration Pythagoras had actually used, we may never know.[9] In any event, the proof heralded a fundamental change in the way we think about mathematics: no longer could one simply discover a new relation between mathematical objects; one had to *prove* it by a logically consistent argument. This change marked the transition from the empirical nature of pre-Greek mathematics to the deductive discipline it is today.

❖   ❖   ❖

The 45-45-90–degree triangle naturally led the Pythagoreans to the following problem: given a square of unit side, find the length of its diagonal (fig. 2.8). Denoting this length by d and using the Pythagorean theorem, we have $d^2 = 1^2 + 1^2 = 2$, so that (in modern notation) $d = \sqrt{2}$. But what kind of number is $\sqrt{2}$? A simple geometric construction, using straightedge and compass (fig. 2.9), shows that the value of $\sqrt{2}$ is somewhere between 1 and 2, so it must be a fraction—a ratio of two integers. But no matter how hard the Pythagoreans tried to find this ratio, they failed: many ratios came close, but

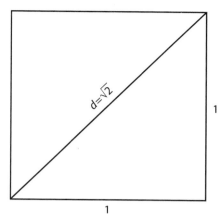

Figure 2.8. The diagonal of a square

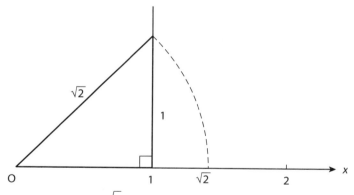

Figure 2.9. Construction of $\sqrt{2}$ on the number line

none was exactly equal to $\sqrt{2}$. Thus the name *irrational*, meaning "not rational" (in the mathematical sense). Note, however, the double meaning of the word: it also means "not governed by reason," which was exactly how the Greeks reacted when they discovered that $\sqrt{2}$ is not a ratio of integers.

The exact circumstances of this discovery, like all others made by the Pythagoreans, are shrouded in mystery; we do not know if it was Pythagoras himself or one of his disciples who should get the credit, nor what proof was used (in Appendix D we give one proof). In any case, the discovery left them bewildered, for here was a geometric quantity—the diagonal of a square—whose length could not be expressed as a ratio of two integers. This at once shattered their belief in the supremacy of rational numbers, and it had profound repercussions. Not knowing how to deal with this new kind of magnitude, the Pythagoreans refused to consider it a number at all, in effect regarding the diagonal of a square as a numberless quantity. Thus a rift was opened between geometry and arithmetic, and by extension, between geometry and algebra. This rift would persist for the next two thousand years and was to become a major impediment to the further development of mathematics; it was only in the seventeenth century, with the invention of analytic (coordinate) geometry by René Descartes, that the divide was finally bridged.

According to legend, the Pythagoreans were so shaken by the discovery that $\sqrt{2}$ is irrational that they vowed to keep it a closely guarded secret, perhaps fearing it might have adverse effects on the populace. But one of them, a person by the name of Hippasus, resolved to reveal the discovery to the world. Enraged by this breach of loyalty, his fellows cast him overboard the boat they were sailing, and his body rests to this day at the bottom of the Mediterranean.

❖   ❖   ❖

The Pythagorean legacy lasted well over two thousand years, and to some degree continues to the present. The Pythagorean obsession with number symbolism and number mysticism influenced countless writers, artists, and thinkers. The relations between mathematics and music, first discovered by Pythagoras, had repercussions in Renaissance Europe, where cathedrals were designed according to the musical proportions 2:1, 3:2, and 4:3. An illustration from a book published in 1617 (fig. 2.10) shows God's hand tuning a giant monochord, a string stretched along a sounding board on which the planetary orbits are superimposed over the intervals of the musical scale. The phrase "music of the spheres" became an inspiration for many Renaissance scientists. One of the last Pythagoreans was the astronomer Johannes Kepler (1570–1631). At once a scientist of the highest rank and a diehard mystic, Kepler spent— some would say wasted—a good thirty years of his life seeking to discover the laws of planetary motion in musical harmony. Kepler believed that each planet plays a certain tune according to its distance from the Sun—the closer the planet to the Sun, the higher the notes (fig. 2.11). Yet it was Kepler who put to rest the old Greek circular orbits, replacing them by ellipses and thereby setting astronomy on its modern course.

Ghosts of the Pythagorean philosophy linger on to this day. In the wake of Einstein's masterful formulation of his general theory of relativity, the idea that simplicity and symmetry play a role in the laws of nature once again gained ground. The physicist Paul A. M. Dirac (1902–1984) set the tone in his famous adage, "It is more important to have beauty in one's equations than to have them fit experiment."[10] At present, a number of distinguished theoreticians are devoting their careers to an esoteric subject called string theory, which purports to explain everything in the universe—from the Big Bang to the inner workings of subatomic particles—in terms of strings vibrating in

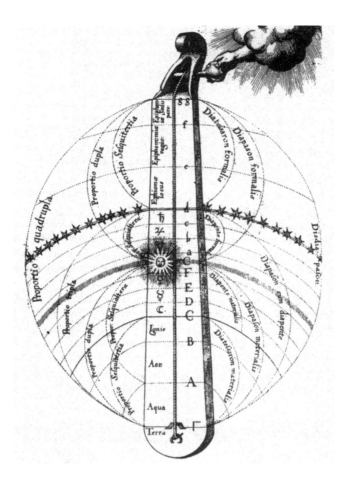

Figure 2.10. God's hand tuning the universal monochord

Figure 2.11. "The Harmonics of the Planets," from Johannes Kepler's *Harmonia Mundi* (1619)

an eleven-dimensional space. Vibrating strings? Pythagoras would have been delighted.

Notes and Sources

1. The details of Pythagoras's life given here are partly based on Smith. In particular, see vol. 1, pp. 69–77, and vol. 2, pp. 288–290.

2. Our main source of information on Pythagoras's work comes from the *Eudemian Summary* of Proclus, a work that contains a commentary on Book I of the *Elements* and a historical outline of Greek geometry up until Euclid's time; this summary was based on fragments of an earlier work by Eudemus (fl. ca. 335 BCE), a student of Aristotle. Although Proclus (412–485 CE) lived a thousand years after Pythagoras, he may still have had access to some original writings of his predecessors.

3. As of 2005, the largest perfect number was $2^{25,964,950} \times (2^{25,964,951} - 1)$, a 15,632,458-digit monster discovered in that year by Dr. Martin Nowak, a German eye surgeon, after a fifty-day calculation using a 2.4 GHz Pentium 4 computer. Euclid, in the third century BCE, proved that if $2^n - 1$ is prime, then $2^{n-1}(2^n - 1)$ is a perfect number (and by necessity, even). Some two thousand years later, in 1770, Leonhard Euler proved that *every* even perfect number is of the form $2^{n-1}(2^n - 1)$. Although nothing in theory precludes a perfect number from being odd, not a single odd perfect number has yet been found.

Primes of the form $2^n - 1$, where n itself is prime, are known as *Mersenne primes*, named after Marin Mersenne (1588–1648), a French friar of the order of Minims and a freelance mathematician. The theorems of Euclid and Euler mentioned above mean that every Mersenne prime generates a perfect number; hence their histories are closely linked. The first four values of n to yield Mersenne primes are 2, 3, 5, and 7, resulting in the primes 3, 7, 31, and 127. Not every prime n yields a Mersenne prime; for $n = 11$ we get $2^{11} - 1 = 2,047 = 23 \times 89$, a composite number. Thus, the requirement that n be prime is a *necessary*, but not *sufficient*, condition for $2^n - 1$ to be prime.

4. Because of the triangular shape of the pattern of dots, the sum of the first n integers is called a *triangular number*. The first ten triangular numbers are 1, 3, 6, 10, 15, 21, 28, 36, 45, and 55.

Nowadays we would regard this sum as a special case of an arithmetic progression, where each member is obtained from its predecessor by adding a fixed number, in this case 1. There is a story about the great German mathematician Carl Friedrich Gauss (1777–1855), who as a ten-year-old boy was asked by his teacher to add up the first one hundred integers. To the teacher's amazement, Gauss almost immediately came up with the correct answer, 5,050. Gauss explained that he simply wrote the required sum twice, once as $1 + 2 + 3 + \ldots + 98 + 99 + 100$, and again as $100 + 99 + 98 \ldots + 3 + 2 + 1$, and then added the two rows vertically. Each pair adds up to 101, and there are hundred such pairs, making the sum of the two rows $100 \times 101 = 10,100$. The required sum is one half of this, or 5,050.

For more about figurative numbers, see Eves, pp. 78–80 and 94.

5. We should mention, however, that the Greeks did not regard the quantities a and b as variables, as we do today, but as fixed geometric magnitudes represented by line segments. The notion of a variable quantity was foreign to them, a fact that prevented

them from transforming their geometric algebra into the powerful tool that symbolic algebra would become many centuries later.

6. The account, however, is highly doubtful: bound by their ascetic code of conduct, the Pythagoreans abhorred the slaughter of animals. Still, the story, according to one nineteenth-century German poet, had a chilling effect on the community of oxen, who from that moment on trembled in fear at the mere mention of a new mathematical discovery (see page 46).

7. For more on Pythagorean triples, see Appendix B. See also Eves, pp. 81–82 and 97–98.

8. Much scholarly debate has evolved around the question of how the Pythagoreans proved the theorem, but we are none the wiser for it. See *Euclid: The Elements*, translated, with introduction and commentary, by Sir Thomas Heath (New York: Dover, 1956), vol. 1, pp. 350–356.

9. The extent to which scholars are unsure as to the kind of proof Pythagoras had used can be seen from two diametrically opposite opinions by two highly regarded historians of mathematics, Howard Eves and Sir Thomas Heath. Eves says, "It is generally felt that it [Pythagoras's proof] probably was a dissection type of proof" (Eves, p. 81), while Heath is of the opinion that "there is difficulty in supposing that Pythagoras used a general proof of this kind [by dissection] . . . ; it has no specifically Greek colouring but rather recalls the Indian method" (Heath, *Euclid*, vol. 1, p. 355). Heath does not, however, preclude the possibility that Pythagoras used such a proof in the case of triangles with *rational* sides [e.g., the (3, 4, 5) triangle].

10. Quoted from his article, "The Evolution of the Physicist's Picture of Nature," *Scientific American*, May 1963.

Euclid's *Elements*

No other proposition of geometry has exerted so
much influence on so many branches of
mathematics as has the simple quadratic formula
known as the Pythagorean theorem. Indeed, much
of the history of classical mathematics, and of
modern mathematics, too, could be written
around that proposition.
    —Tobias Dantzig, *The Bequest of the Greeks*, p. 95

The Pythagorean school soon acquired the reputation of an exclusive, aristocratic club—today we would say an elitist group—and it was not long before its members drew ire from their fellow citizens. They were harassed, their meeting places were destroyed, and Pythagoras himself was either forced to flee or was killed (like the rest of his life, the circumstances of his death are uncertain; he reportedly was nearing eighty when he died). Yet far from being finished off, the Pythagorean legacy had just begun.

Meanwhile, great political changes were reshaping the ancient world. The Persian Empire was rising to prominence and soon replaced Babylon as the dominant power east of the Mediterranean. In the year 546 BCE Persia conquered the Ionian cities and their colonies in Asia Minor. In 499, Athens led a failed revolt against the Persians. Seeking revenge, King Darius sent a huge armada to attack mainland Greece, but his fleet was destroyed in a storm. In 490 the Athenians defeated the Persian army at Marathon, and Athens asserted its political dominance over the other Greek city-states.

There followed half a century of peace, during which Athens flourished as a center of democracy and learning. The dispersed Pythagoreans found refuge there, and great thinkers like Pericles, Socrates, Anaxagoras, Zeno, and Parmenides made Athens their home. But the period of calm ended with the Peloponnesian War (431–404), during which Athens was crushed militarily by Sparta and decimated by plague. Then in 371 Sparta itself was defeated at the hands of rebellious city-states. The center of learning moved to Tarentum (now Italy), where the Pythagoreans started afresh under the leadership of Archytas.

But Athens slowly regained its leading role. Its moment of glory came in 387 when the great philosopher Plato (ca. 427–347) founded the Academy of Athens, which was to dominate Greek intellectual life for the next thousand years. Though not a mathematician himself, Plato's main contribution to mathematics was his recognition of its importance to learning in general, to logical thinking, and, ultimately, to a healthy democracy (how true even today!). His motto, inscribed over the entrance to the Academy, became immortal: "Let no one unversed in geometry enter here."

The next great political upheaval came in 338 when Greece, following a military campaign by King Philip, became part of the Macedonian nation. Two years later, his son, Alexander the Great (356–323) ascended to the throne, and within ten years expanded the Greek empire to encompass nearly the entire ancient world, from the gates of India in the east to the Pillars of Hercules (the Strait of Gibraltar) in the west.

In 332 Alexander founded a new city at the western extremity of the Nile Delta in Egypt and named it after himself. Alexandria soon became the commercial and intellectual capital of the Hellenistic Empire. By 300, its population grew to half a million, and it boasted the most magnificent buildings in the ancient world. At the entrance to its large harbor stood an imposing lighthouse, 300 feet tall, whose fiery torch could be seen from a distance of 70 miles—one of the seven wonders of antiquity.

But just nine years later, the course of history changed again. In 323 Alexander died prematurely at the age of thirty-three. The Hellenistic Empire split into separate parts, divided by political rivalries yet kept together by the cultural heritage of Alexander. Egypt fell to the rule of the Ptolemy dynasty. Ptolemy I began his reign in 306, choosing Alexandria as his capital and founding a school there that was to become the crown jewel of the ancient world. It had all the trappings of a modern university campus: magnificent buildings, gardens and dormitories, and a museum. Its famed library boasted over half a million books (in the form of papyrus scrolls), which were acquired—sometimes illegally and by coercion—anywhere they could be found. Scholars from near and afar came to Alexandria for extended periods of study. Thanks to them, Greek culture became the dominant culture of the ancient world, and its language, the lingua franca of the day.[1]

It is at this juncture that Euclid entered the scene. As with Pythagoras, almost nothing is known about his life; even his year and place of birth are uncertain, but it seems likely he was raised and educated in Athens. He then settled in Alexandria and became head of the mathematics department at the university (by some accounts, he was the chief librarian of its library). His famous quip, in reply to Ptolemy's request for a shortcut to studying the subject, "There is no royal road to geometry," may have actually been someone else's.

There is also the story that in response to a student who wanted to know what geometry was good for, Euclid gave him a penny, "since he must make gain from what he learns."

Euclid wrote several books on mathematics and optics, some of which have survived through Arabic translations. But by far his most influential work was the *Elements*. Written in thirteen "books" (today we would say parts), it is a compilation of the state of mathematics as it was known in his time. Its terse, rigorous style—definitions, axioms, theorems, and proofs—is a model of mathematical writing to this day. The thirteen books contain 465 theorems covering geometry, number theory, and primitive algebra ("primitive" in the sense that algebraic formulas are derived geometrically rather than symbolically). It is not known which of these theorems, if any, were discovered by Euclid himself. He took upon himself the role of editor in chief of a huge project, assembling into a single, logically arranged structure the large body of mathematics that had been acquired since Pythagoras. As such he left his own contributions, if there were any, entirely without credit.

It is instructive to compare The *Elements* to a modern mathematics textbook. You will not find here the usual preface and introduction, a foreword to the student and a foreword to the instructor, exercises with answers, appendixes, a bibliography, and an index. Nor will you find words of praise about the merits of the book over its competitors. Right from the first page—indeed, the first sentence—it is down to business. This terse style, and the steely logic of the arguments themselves, was exactly what attracted the work to many a great thinker: Descartes, Newton, and numerous other scientists got their introduction to mathematics by studying Euclid, usually on their own and at a young age.

Certainly no book has had a greater impact on mathematics than the *Elements*. The work has been translated into practically every language and reissued in numerous editions (by some accounts, it enjoyed the second largest number of editions after the Bible). Moreover, few works have had a greater number of written commentaries, and commentaries on the commentaries; a modern edition will have perhaps twenty pages of commentary for each page of original text (I will have a word later on what is meant by "original" in this case). In this, and in its concise, matter-of-fact style, the *Elements* can perhaps best be compared with the Talmud, the codifying of Jewish law completed in the fourth century CE.

The *Elements* opens with twenty-three definitions of fundamental concepts, such as a point ("that which has no part"), a line (a "breadthless length"), a straight line ("a line which lies evenly with the points on itself"), and a plane angle ("the inclination to one another of two lines in a plane which meet one another and do not lie in a straight line").[2] These are followed by ten statements that Euclid regarded as self-evident, as so clear and unquestionable as to obviate the need for proof. These statements—today we call them *axioms*—are divided into two groups: the first five deal with geometric concepts and are called "postulates;" the remaining five are arithmetic in nature and are called "common notions." We list them here:

Postulates

1. To draw a straight line from any point to any point.
2. To produce a finite straight line continuously in a straight line.
3. To describe a circle with a centre and distance.
4. That all right angles are equal to one another.
5. That, if a straight line falling on two straight lines makes the interior angles on the same side less than two right angles, the two straight lines, if produced indefinitely, meet on that side on which are the angles less than the two right angles.

(The phrases "to draw," "to produce," and "to describe" are to be interpreted as "one can draw," etc.)

Common Notions

1. Things which are equal to the same thing are also equal to one another.
2. If equals are added to equals, the wholes are equal.
3. If equals are subtracted from equals, the remainders are equal.
4. Things which coincide with one another are equal to one another.
5. The whole is greater than the part.

For the next two thousand years, these ten axioms, and the edifice of 465 theorems that Euclid derived from them, were accepted as the absolute, infallible truth, handed down to us by divine authority; indeed, they were sometimes referred to as the Ten Commandments of Geometry. Moreover, the geometry based on them—Euclidean geometry—was regarded as the *only* geometry possible. It was not until the eighteenth century that doubts began to be raised about the absolute validity of these axioms, and in particular, of the fifth postulate, known as the *parallel postulate*. But it would be another hundred years before mathematicians realized that Euclid's geometry is only one of many possible geometries. The implications were profound and would shake mathematics to its core. This story, though, will have to wait until chapter 12.

Immediately following the ten axioms, without a single further word, comes the first of the forty-eight theorems—called "propositions" by Euclid—that make up Book I of the *Elements*: how to construct an equilateral triangle when its side is given. This is followed by the basic geometric constructions with straightedge and compass, the same constructions every child learns in grade school to this day: how to copy a line segment (that is, move it to a new position in the plane), how to drop a perpendicular to a line from a point outside it, how to bisect an angle, and so on. Here too we find the three familiar congruency theorems (*SAS*, *SSA*, and *SSS*, where *S* stands for "side" and *A* for "angle") and the theorem about the sum of the angles in a triangle equaling two right angles (there is no mention of 180°). Then, near the end of Book I, comes

Proposition 47
In right-angled triangles the square on the side subtending the right angle is equal to the squares on the sides containing the right angle.

That is to say, the square on the hypotenuse is equal to the sum of the squares on the two sides: the Pythagorean theorem. Not once in the *Elements* is there a name of a person associated with a particular proposition, not even that of Pythagoras.

Before he could prove it, Euclid needed a preliminary theorem, listed as

Proposition 38
Triangles which are on equal bases and in the same parallels are equal to one another.

That is to say, triangles with the same base and a top vertex that lies on a line parallel to the base have the same area.

Proof
In figure 3.1, let the triangles *ABC* and *DEF* have equal bases *BC* and *EF* (in the figure they are drawn along the same line). The vertices *A* and *D* are on a line parallel to *BC* and *EF*. We extend *AD* to points *G* and *H*, where *GB* is parallel to *AC* and *HF* is parallel to *DE*. Then the figures *GBCA* and *DEFH* are parallelograms with the same area, for they have equal bases *BC* and *EF* and lie between the same parallels *BF* and *GH*. Now the area of triangle *ABC* is half that of parallelogram *GBCA*, and the area of triangle *DEF* is half that of parallelogram *DEFH*. Therefore the two triangles have the same area.— QED.[3]

(A modern proof would simply make use of the formula for the area of a triangle, area = (base × height)/2, and proceed like this: Let the fixed base *BC* have length *a*, and let the top vertex *A* move on a line parallel to *BC*. Regardless of the position of *A* on the line, its vertical distance from *BC* is constant, say *h*. Thus the area of every triangle fulfilling the given requirements is $\frac{ah}{2}$ = constant.)

Now Euclid was ready to prove the main theorem. As a first step he proved a lemma (a preliminary result): In a right triangle, the square on a side has the same area as the rectangle formed by the hypotenuse and the perpendicular projection of the same side on the hypotenuse.

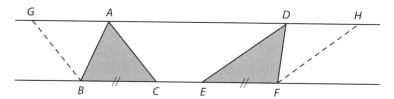

Figure 3.1. Proposition I 38: Triangles *ABC* and *DEF* have the same area

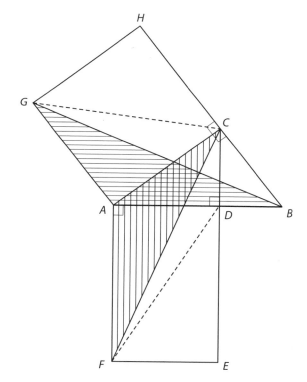

Figure 3.2. Lemma for proving Proposition I 47

To make sense of this, we turn to figure 3.2. Let the right triangle be *ABC*, with the right angle at *C*. On side *AC* we build a square *ACHG*, so ∠*ACH* is a right angle. But ∠*ACB* is also a right angle, sharing side *AC* with ∠*ACH*. Thus ∠*HCB* = ∠*HCA* + ∠*ACB* = two right angles, and therefore *HC* is an extension of *BC*. Let the perpendicular projection of *AC* on the hypotenuse be *AD*. Construct *AF* equal to *AB* and perpendicular to it, and consider the triangles *BAG* and *FAC* (both shaded in the figure). We have *AF* = *AB* and *AC* = *AG*. Moreover, ∠*BAG* = ∠*FAC*, as each consists of a right angle and the common angle

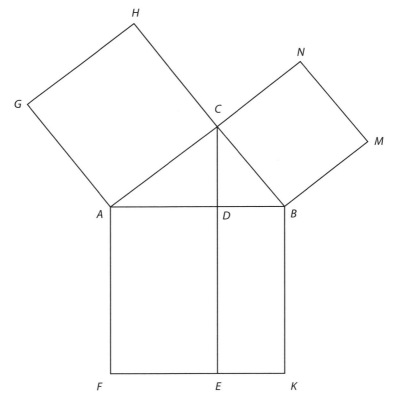

Figure 3.3. Proof of Proposition I 47

BAC. Thus Δ*BAG* and Δ*FAC* are congruent (the *SAS* case) and therefore have the same area. But by the preliminary theorem, Δ*BAG* has the same area as Δ*CAG* (where *CG* is the diagonal of the square *ACHG*), because vertices *B* and *C* lie on a line parallel to the base *AG*. Similarly, Δ*FAC* has the same area as Δ*FAD* (where *FD* is the diagonal of rectangle *AFED*). But Δ*CAG* has half the area of square *ACHG*, and Δ*FAD* has half the area of rectangle *AFED*. Tying all this together, we finally get the result

area ACHG = area AFED.

Now (if you have persevered up to this point) we are almost there. What is true for one side of the right triangle is, of course, true for the other side as well. Referring to figure 3.3, we thus have

area ACHG = area AFED

area BCNM = area BKED.

Adding the two equations, we get

area *ACHG* + area *BCNM* = area *AFKB*,

which is the Pythagorean theorem.—QED.[4]

❖   ❖   ❖

Now this must surely be among the most difficult proofs a beginning geometry student is likely to encounter. But it was this proof that generation after generation of schoolchildren had to struggle with, including this author (fig. 3.4).
The philosopher Arthur Schopenhauer is said to have protested that this proof,
rather than instruct the student, could easily overwhelm him: "Lines are drawn,
we know not why, and it afterwards appears they were traps which close unexpectedly and take prisoner the assent of the astonished reader."[5]

So the question naturally arises, why did Euclid choose this particular proof,
when there are so many simpler ones—*much* simpler ones? There are two probable answers. First, most of these other proofs depend on dissecting a right triangle into smaller, similar triangles, and then using the laws of proportion to derive
the equation $a^2 + b^2 = c^2$. But the theory of proportion is not introduced in the
Elements until Book V, and the laws of similarity not until Book VI, so Euclid
could not have used them at this stage, or else he would be following a "circular
argument," among the worst sins a mathematician could possibly commit.

Of course, Euclid could have used the "Chinese" proof (see p. 25), which
certainly seems simpler than I 47. This proof, involving the dissection of a
square, is a "dynamic" proof, based on the fact that a planar figure does not
change its area when moved around as a rigid body. But such a demonstration,
depending on notions from the physical world, was anathema to Euclid, who
insisted that the validity of every statement should be established by deductive
reasoning alone. This essentially ruled out any dissection, "cut-and-paste"
kind of proof.

This brings us to the second reason. In his classic book *The Bequest of the
Greeks*, Tobias Dantzig claimed that Euclid's proof "interprets the Pythagorean
theorem not as a *metric* relation between the sides of a right triangle, but as a
property of the *squares* erected on these sides. This literal interpretation of the
theorem restricts the proof to *areal* [area-related] *equivalence*."[6] We must remember again that the Greeks interpreted all arithmetic operations in a geometric context. A number was regarded as the length of a line segment; the sum of
two numbers, as the combined length of two segments laid out end-to-end; and a
product of two numbers, as the area of a rectangle with the corresponding segments as sides. As a special case, the square of a number *a* was interpreted as the
area of a square of side *a* (which is why the quantity a^2 is called "*a* squared").
So it was natural for the Greeks to regard the Pythagorean theorem as a relation
between areas—indeed, it was the *only* way they could think of it. Viewed in this
light, it made perfect sense for Euclid to prove the theorem the way he did.

CA על היתר. על הנצב AC נבנה רבוע ACHG. הצלע CH היא המשך של AC.
נעביר את CF ו BG. במשולשים CAF ו GAB

$$AC = AG \qquad AF = AB$$

כל אחת מהזויות CAF ו GAB מורכבת מהזוית CAB ומזוית ישרה, לכן הן
שוות. המשולשים CAF GAB חופפים, איפוא,
ע"י משפט החפיפה הראשון[1]). נעביר את
האלכסונים DF ו CG. המשולשים DAF ו CAF
שוים בשטחם, מפני שיש להם צלע משותפת AF
וקדקדיהם D ו C נמצאים על ישר המקביל לצלע
הזאת. המשולשים CAG ו BAG שוים בשטחם,
מפני שיש להם צלע משותפת AG וקדקדיהם C
ו B נמצאים על ישר המקביל לצלע הזאת. לכן,
המשולשים DAF ו CAG שוים בשטחם. חמשולש
הראשון הוא חצי המלבן DAFE והמשולש השני

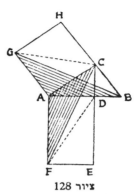

ציור 128

הוא חצי הרבוע CAGH; נמצא כי המלבן והרבוע שוים בשטחם.

§163. משפט. במשולש ישר־זוית, הרבוע הבנוי על היתר
שוה בשטחו לסכום הרבועים הבנויים על הנצבים.

הוכחה. נתון משולש ABC [ציור 129] עם זוית ישרה C. נבנה רבוע
על כל אחת מצלעותיו. נעביר את CE באופן
שיהיה מאונך ל AB. המלבן ADEF שוה בשטח
לרבוע הבנוי על הנצב AC. המלבן BDEL שוה
בשטחו לרבוע הבנוי על הנצב BC. לכן, הרבוע
ABLF שוה בשטחו לסכום הרבועים הבנויים על
הנצבים.

תוצאה. במשולש ישר־זוית, הרבוע
הבנוי על אחד הנצבים שוה בשטחו
להפרש שבין הרבועים הבנויים על
היתר ועל הנצב השני.

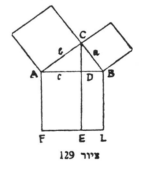

ציור 129

אם נסמן ב c את היתר, ב a ו b — את הנצבים, יהיה:

$$a^2 + b^2 = c^2 \qquad a^2 = c^2 - b^2$$

---

1) בכדי לחפות את המשולש GAB במשולש CAF, צריך להסב את זה האחרון
על חנקודה A זוית של 90°.

Figure 3.4. The Pythagorean theorem in the author's high school geometry textbook

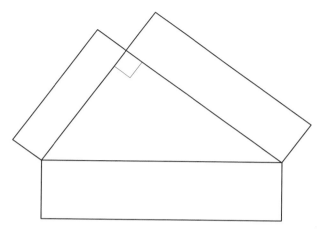

Figure 3.5. Euclid's drawing for Proposition VI 31

Nevertheless, he must have recognized the difficulties this proof presented to the reader, for in Book VI Euclid offered a second proof, this one based on similarity. Proposition VI 31 says,

In right-angled triangles the figure on the side subtending the right angle is equal to the similar and similarly described figures on the sides containing the right angle.

This is an almost verbatim repetition of Proposition I 47, except that "square" is replaced by "figure," and "squares" by "similar and similarly described figures." A drawing accompanying the theorem in the *Elements* (fig. 3.5) shows three similar rectangles, but the "described figures" could be *any* similarly constructed figures; they don't even have to be polygons. In this sense, VI 31 is actually a more general form of the Pythagorean theorem than I 47. An outline of Euclid's proof follows.[7]

In figure 3.6, let the right triangle be *ABC*, with the right angle at *C*. As with the proof of I 47, we drop the perpendicular *CD* from *C* to the hypotenuse *AB*. Because $AB \perp CD$ and $AC \perp CB$, we have $\angle DAC = \angle DCB$. Therefore, triangles *ADC* and *CDB* are similar to each other and to the whole triangle *ACB*.[8] It follows that $AB/AC = AC/AD$ and $AB/BC = BC/BD$, from which we get, by cross-multiplying,

$$AC^2 = AB \times AD \ \text{ and } \ BC^2 = AB \times BD.$$

Adding these equations, we have

$$AC^2 + BC^2 = AB \times AD + AB \times BD = AB \times (AD + BD).$$

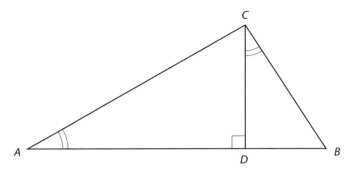

Figure 3.6. Proof of Proposition VI 31

But $AD + BD = AB$, so we finally get

$$AC^2 + BC^2 = AB \times AB = AB^2. \qquad\qquad \text{QED}$$

The commentator Proclus (ca. 412–485 CE), whose *Eudemian Summary* is our chief source on early Greek mathematics, credited Euclid as the originator of this proof; according to Proclus, it was the *only* proposition in the *Elements* proved by Euclid himself.[9] At any rate, it is this second proof that has been favored by modern writers; as one seventeenth-century author said, "I've thought a lot about it, but have never managed the proof by any other way than by proportions [i.e., by similarity]."[10]

As for the proof of I 47, Proclus attributed it to Euclid's predecessor, Eudoxus of Cnidus (ca. 408–ca. 355 BCE). Eudoxus was the discoverer of the *method of exhaustion*, a powerful mathematical procedure that two centuries later would be used to great effect by Archimedes; he is also believed to be the originator of the theory of proportion, the subject of Book V of Euclid. Of Pythagoras's original proof of the theorem named after him—if indeed he had a valid, general proof—we have no trace whatsoever. Considering the still primitive stage of mathematics in his time, it is highly unlikely that either I 47 or VI 31 originated with him (however, we will have more to say about this in chapter 5). As with so much else about early Greek mathematics, whoever should be credited for first proving the most celebrated theorem in mathematics remains unknown.

I 47 is the next-to-last proposition of Book I. The concluding proposition, number 48, is little known and rarely mentioned in geometry texts:

> *If in a triangle the square on one of the sides be equal to the squares on the remaining two sides, the angle contained by the remaining two sides is right.*

This is the *converse* of the Pythagorean theorem; in essence it says that a right triangle is the *only* triangle for which the equation $a^2 + b^2 = c^2$ holds true.

The proof is simple; we give it here in modern notation. Let the triangle have sides a, b, and c, with $c^2 = a^2 + b^2$. Construct a *right* triangle with sides equal to a and b, and let its hypotenuse be d. By I 47, $d^2 = a^2 + b^2$. But $a^2 + b^2 = c^2$ (given), so $d^2 = c^2$; that is, the squares built on c and on d have the same area, and are therefore congruent. It follows that $d = c$. But then the given triangle and the one just constructed have all of their three sides equal, and are thus congruent (the *SSS* congruency case). And since triangle (a, b, d) is a right triangle, so is triangle (a, b, c), with the right angle opposite of side c.—QED.[11]

With this, Euclid closed Book I. Although Pythagoras was never mentioned by name, Euclid must have felt that the Pythagorean theorem would be an apt conclusion to the first book of his work, an implicit tribute to the great master. Euclid would make frequent use of it in the rest of the *Elements*. And his two proofs, I 47 and VI 31, would over the coming centuries multiply a hundredfold; well over four hundred are known today. With Euclid, the legacy of the Pythagorean theorem had just begun.

Notes and Sources

1. This brief sketch of Greek history is based on Eves, pp. 105–108 and 140–141. For more on the library of Alexandria, see Lionel Casson, *Libraries in the Ancient World* (New Haven, Conn., and London, U.K.: Yale University Press, 2001), chapter 3.

2. All definitions and axioms given here are from the translation, with introduction and commentary, by Sir Thomas Heath (in 3 vols.; New York: Dover, 1956).

3. QED is an acronym of the Latin *quod erat demonstrandum*—"that which was to be demonstrated." In modern American textbooks this is usually replaced by a small square. Older books often use the symbol ∴ .

4. Eves (pp. 155–156) suggests that Euclid's proof can be made into a "dynamic proof, . . . , wherein the square on the hypotenuse is continuously transformed into the sum of the squares on the legs of the right triangle." See figure 3.7.

5. As quoted in J. L. Heilbron, *Geometry Civilized: History, Culture, and Technique* (Oxford: Oxford University Press, 1998), p. 147. Heilbron in turn quotes Florian Cajori, *Mathematics in Liberal Education* (Boston: Christopher, 1928).

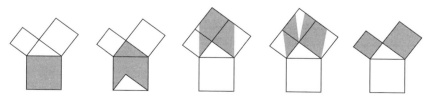

Figure 3.7. A "dynamic proof"

6. (New York: Charles Scribner's Sons, 1955), p. 97. The italics in the quotation are in the original.

7. See, however, Heath, vol. 3, pp. 269–270, for a discussion of some subtleties that Euclid may have omitted.

8. We write *ACB* rather than *ABC* in order to keep the correct order of letters in the similarity relations; the line segments themselves, however, are *nondirected*; that is, $AB = BA$, etc.

9. On this subject, see Dantzig, *Bequest of the Greeks*, pp. 97–99; Heath, vol. 2, pp. 269–270.

10. Heilbron, *Geometry Civilized*, p. 147.

11. We might be tempted to take the square root of both sides of the equation $d^2 = c^2$ and conclude that, since c and d denote length and are therefore positive, $d = c$. But as already mentioned, this is not the way Euclid proceeded; he followed a strictly geometric approach based on area relations.

The Pythagorean Theorem in Art, Poetry, and Prose

The Square of the Hy-Pot-E-Nuse . . .
—Theme From a musical ditty by Saul Chaplin,
with lyrics by Johnny Mercer

It has been known by many names: *Euclid 1, 47*, because it is listed as Proposition 47 of Book I of Euclid's *Elements*; the *Windmill*, since its characteristic configuration resembles the three sails of a mill; the *Bride's Chair*, for reasons known only to the person who proposed it;[1] *Dulcarnon* ("two horned," the shape of the Franciscan's hood);[2] or simply the *Hypotenuse theorem*. Renaissance mathematician Luca Pacioli (1445–1509) called it the *Goose foot and the Peacock's tail*,[3] while to the Chinese it was known as the *Kou-ku theorem*. Perhaps strangest of all is *pons asinorum*, the "Bridge of Asses"; this name is usually associated with a different theorem—the one asserting the equality of the two base angles in an isosceles triangle (Euclid I 5)—but the French reserve it for the characteristic figure of the Pythagorean theorem.[4]

Lewis Carroll (1832–1898), the author of *Alice's Adventures in Wonderland* (1865) and *Through the Looking-Glass* (1871), is one of the most beloved writers of children's tales. Less well known is the fact that he was also a mathematician whose real name was Charles Lutwidge Dodgson. But even in his mathematical writing he could not suppress his inexhaustible sense for pun and humor. Here is an excerpt from his book *A New Theory of Parallels* (London, 1895):

> But neither thirty years, nor thirty centuries, affect the clearness, or the charm, of Geometrical truth. Such a theorem as, "the square of the hypotenuse of a right-angled triangle is equal to the sum of the squares of the sides" is as dazzlingly beautiful now as it was in the day when Pythagoras first discovered it, and celebrated its advent, it is said, by sacrificing a hecatomb of oxen—a method of doing honor to Science that has always seemed to me *slightly* exaggerated

and uncalled-for. One can imagine oneself, even in these degenerate days, marking the epoch of some brilliant scientific discovery by inviting a convivial friend or two, to join one in a beefsteake and a bottle of wine. But a *hecatomb* of oxen! It would produce a quite inconvenient supply of beef.[5]

On the same subject, the German writer Karl Ludwig Börne (1786–1837) had this to say: "After Pythagoras discovered his fundamental theorem he sacrificed a hecatomb of oxen. Since that time all dunces [*Ochsen*; in the German vernacular a dunce or blockhead is called an ox] tremble whenever a new truth is discovered."[6]

The German poet and botanist Adelbert von Chamisso (1781?–1838) wrote a little ditty on the same theme, which I give here in translation:

> Truth lasts throughout eternity,
> When once the stupid world its light discerns:
> The theorem, coupled with Pythagoras' name,
> Holds true today, as't did in olden times.
>
> A splendid sacrifice Pythagoras brought
> The gods, who blessed him with this ray divine;
> A great burnt offering of a hundred kine,
> Proclaimed afar the sage's gratitude.
>
> Now since that day, all cattle [blockheads] when they scent
> New truth about to see the light of day,
> In frightful bellowing manifest their dismay;
>
> Pythagoras fills them all with terror;
> And powerless to shut out light by error,
> In sheer despair they shut their eyes and tremble.[7]

Jacob Bronowski (1908–1974), the noted physicist, biologist, author, and TV commentator, had this to say in *Science and Human Values*: "To this day, the theorem of Pythagoras remains the most important single theorem in the whole of mathematics. That seems a bold and extraordinary thing to say, yet it is not extravagant; because what Pythagoras established is a fundamental characterisation of the space in which we move, and it is the first time that it is translated into numbers. . . . In fact, the numbers that compose right-angled triangles have been proposed as messages which we might send out to planets in other star systems as a test for the existence of rational life there."[8]

In slightly less reverent words, American physicist and Nobel laureate Leon Lederman (1922–) described Pythagoras as "the first cosmic guy. It was he (and not Carl Sagan) who coined the word *kosmos* to refer to everything in our universe, from human beings to the earth to the whirling stars overhead. *Kosmos* is an untranslatable Greek word that denotes the qualities of order and beauty. The universe is a *kosmos*, he said, an ordered whole, and each of us humans is also a *kosmos* (some more than others)."[9]

The following words come from Johannes Kepler (1571–1630), at once the father of modern astronomy and one of the last Pythagoreans: "Geometry has two great treasures: one is the theorem of Pythagoras, the other the division of a line into extreme and mean ratio. The first we may compare to a measure of gold; the second to a precious jewel."[10]

At least one famous writer was turned on to geometry after mastering Euclid's proof of the Pythagorean theorem. This was the English political philosopher Thomas Hobbes (1588–1679). As told by the biographer John Aubrey (1626–1697) in *Brief Lives*:

> He was 40 years old before he looked on Geometry; which happened accidentally. Being in a Gentlemen's Library, *Euclid's Elements* lay open, and 'twas the 47 *El. libri I.* He read the Propositions. By G——, sayd he (he would now and then sweare an emphaticall Oath by way of emphasis) *this is impossible!* So he reads the demonstration of it, which referred him back to such a Proposition; which Proposition he read. That referred him back to another, which he also read. *Et sic deinceps* [and so on] that at last he was demonstratively convinced of that trueth. This made him in love with Geometry.[11]

The famous theorem got its share of recognition on stage as well. In William S. Gilbert and Arthur S. Sullivan's *The Pirates of Penzance*, the Major General gleefully reminds us that

> I'm very well acquainted, too, with matters mathematical,
> I understand equations, both the simple and quadratical,
> About binomial Theorem I'm teeming with a lot o' news,
> With many cheerful facts about the square of the hypotenuse.

In the seventeenth century, the Renaissance, with its ideals of humanism and universalism, reached its high point. Poets, artists, and philosophers reveled in the new vistas opened up by the discoveries of Galileo and Newton. Mathematics and science, no longer the exclusive domain of scholars, found their way into the humanities, where they were embraced with relish. It became fashionable for artists to adorn their work with various geometric objects that often carried allegorical meaning. French artist Laurent de la Hyre (1606–1656) painted a series of canvases to honor the seven Liberal Arts—the ancient *quadrivium* of arithmetic, geometry, music and astronomy, and the medieval *trivium* of grammar, rhetoric, and dialectic—which an educated person was expected to master. One work in this series, *Allegory of Geometry* (1649; see book's cover), is of particular interest to us here. On the left side of the large, 104×218 cm canvas we see a painting standing on an easel; this picture-within-a-picture has several hints to the work of the architect and engineer Gérard Desargues (1593–1662) on perspective. Dominating the scene in the main painting is a reclining woman displaying a parchment on which several geometric figures can be seen. We immediately recognize the figure at the top left: Euclid's proof of the Pythagorean theorem (see p. 37). The figure to its right has been identified as Proposition 9 of Book II of the *Elements*, and the one at the bottom, as Proposition 36 of Book III.[12]

Several countries have commemorated Pythagoras and the Pythagorean theorem on postage stamps, of which we show a selection in plate 1.

Notes and Sources

Note: The song cited in the epigraph was originally sung by Danny Kaye and featured in the film *Merry Andrew* in 1958. For the full score, see Clifton Fadiman, *The Mathematical Magpie* (New York: Simon and Schuster, 1962), pp. 241–244.

1. Smith (vol. 2, pp. 289–290) explains this name as "possibly because the Euclid figure is not unlike the chair which a slave carries on his back and in which the Eastern bride is sometimes transported to the ceremony." According to Smith, the Greeks are said to have used a similar suggestive name, the "theorem of the married women."

2. Vera Sanford, *A Short History of Mathematics* (Cambridge, Mass.: Houghton Mifflin, 1958), p. 272.

3. Ibid.

4. Dan Pedoe, *Geometry and the Liberal Arts* (New York: St. Martin's Press, 1976), p. 153. The French term is *pont aux ânes*. See also *Euclid: The Elements*, Heath's translation, vol. 1, pp. 415–416 (on the origins of the term) and pp. 417–418 (on other popular names for the Pythagorean theorem).

5. Robert Edouard Moritz, *On Mathematics and Mathematicians (Memorabilia Mathematica)* (1914; rpt. New York: Dover, 1942), pp. 307–308.

6. Quoted in Moszkowski, *Die unsterbliche Kiste* (Berlin, 1908); from Moritz, *On Mathematics*, p. 308.

7. *Gedichte*, 1835, trans. from the German; from Moritz, *On Mathematics*, pp. 308–309.

8. Quoted in Dick Teresi, *Lost Discoveries: The Ancient Roots of Modern Science—from the Babylonians to the Maya* (New York: Simon and Schuster, 2002), p. 17.

9. *The God Particle: If the Universe Is the Answer, What Is the Question?* (with Dick Teresi; Boston, Mass.: Houghton Mifflin, 1993), p. 66. The mention of Carl Sagan is in reference to Sagan's Television series *Cosmos*, which in the 1980s brought astronomy to the living rooms of millions of viewers.

10. As quoted by Pedoe, *Geometry*, p. 72. The "division of a line into extreme and mean ratio" refers to the *golden section*, the ratio in which a line segment must be divided if the whole segment is to the larger part as the large part is to the small. This ratio, usually denoted by the Greek letter ϕ, is equal to $(1 + \sqrt{5}) / 2 \sim 1.61803$. Also known as the *divine proportion*, it has a long history and enjoys many interesting properties; see Mario Livio, *The Golden Ratio: The Story of Phi, the World's Most Astonishing Number* (New York: Broadway Books, 2002).

11. Quoted in Stuart Hollingdale, *Makers of Mathematics* (London: Penguin Books, 1991), p. 39.

12. See J. V. Field, *The Invention of Infinity* (Oxford, U.K.: Oxford University Press, 1997), pp. 214–220. *Allegory of Geometry* emerged from a private collection only in 1993 and is now at the Toledo (Ohio) Museum of Art. As of this writing, some doubts have been raised as to whether la Hyre himself painted this work, but no final verdict has so far been reached.

4

Archimedes

> These machines [of war] he [Archimedes] had
> designed and contrived, not as matters of any
> importance, but as mere amusements in geometry.
> —Plutarch, *Life of Marcellus*

After Euclid, the next great name in the history of mathematics is Archimedes (287–212 BCE). Born in Syracuse on the island of Sicily, he is universally regarded as the greatest scientist of the ancient world. Archimedes embodied the image of the pure mathematician par excellence, a person devoted to studying science for its own sake, yet he also applied his discoveries to a wide range of practical matters. Archimedes' numerous mechanical inventions were legendary. Among them were the screw-driven pump, still in use today in some parts of the world, and the mechanical pulley, with which he was able to hoist a huge boat with almost no effort. He also discovered the laws governing floating bodies. Legend has it that when the ruler of Sicily, King Heron, became suspicious that his crown was made of low-grade rather than pure gold, he called upon Archimedes to investigate. Immersing himself and the crown in Syracuse's public bath, Archimedes weighed the volume of water displaced by the crown, from which he concluded that the crown had indeed been forged. Beside himself with excitement, he ran naked in the streets, shouting, "Eureka" (I found it!).

When the Roman fleet besieged Syracuse, Heron called upon Archimedes to devise weapons with which to defend the city. Leaving aside his more esoteric studies, he designed huge cranes that were placed atop the city's walls; when the enemy ships approached, they were plucked from the sea and hoisted into the air, then let fall to their doom. He is also said to have built enormous parabolic mirrors, akin to today's satellite dishes, which could gather the sun's rays and focus them on the enemy ships, setting them ablaze.[1] Such engineering feats spread his reputation through the empire. When the Romans finally breached the city's defenses, their commander, Marcellus, ordered his troops not to harm the great scientist. A soldier found the old sage on the beach, hunched over a figure he had drawn in the sand.

Ignoring the soldier's order to stand up, Archimedes was slain, ending the life of one of history's most illustrious scientists. (Ironically, it is thanks to this event that we know the exact year of his death, a unique case in ancient chronology).

But it is in pure mathematics that Archimedes left his greatest mark. He was the first to determine the area of a parabolic segment, he discovered many of the properties of spirals, and he found that a sphere inscribed in a cylinder has two-thirds of the surface area of the cylinder and two-thirds of its volume. Deeply impressed by this discovery, Archimedes had requested that a figure of a sphere inside a cylinder be engraved on his tombstone, a wish that Marcellus duly honored.

Many of Archimedes' writings survived through later copying and translations, but some are lost, their existence known only from references by later writers. In 1906 one of these lost works was unexpectedly found in a monastery in Istanbul, giving us an invaluable glimpse into the thought process of one of antiquity's greatest minds.[2]

❖ ❖ ❖

In his book, *Measurement of a Circle*, Archimedes showed that the value of π (the ratio of the circumference of a circle to its diameter) is between $3\frac{10}{71}$ and $3\frac{10}{70}$. His idea was to "squeeze" the circle between a series of inscribed and circumscribing polygons of ever more sides. He then found the perimeters of these polygons, from which he got a series of approximations of π of ever increasing accuracy. Although fairly close estimates of π had already been known before, Archimedes was the first to devise an *algorithm*, a procedure that allows us to approximate π *to any desired accuracy*. His method, which we give here in modern notation, makes repeated use of the Pythagorean theorem.

Figure 4.1 shows a circle with center at O and radius 1. Segment AB represents one side of a regular n-gon (a polygon of n equal sides and n equal angles) inscribed in the circle; call its length s_n. Let OC be the perpendicular bisector of AB, and let its extension meet the circle at D. Because D bisects arc AB, it follows that AD and BD are each one side of a regular $2n$-gon; call its length s_{2n}. Archimedes devised a formula from which he could find s_{2n} if s_n was known.

By applying the Pythagorean theorem to right triangle ACD, we have

$$AD^2 = AC^2 + CD^2 = AC^2 + (OD - OC)^2. \tag{1}$$

A second application of the Pythagorean theorem, this time to right triangle ACO, gives us

$$OC = \sqrt{OA^2 - AC^2}.$$

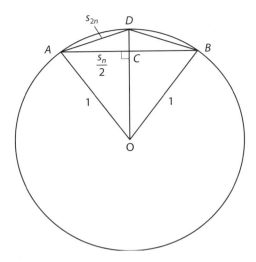

Figure 4.1. A polygon and its inscribed circle

Putting this into equation (1) and recalling that $OD = OA = 1$, $AC = s_n/2$, and $AD = s_{2n}$, we get

$$s_{2n}^2 = (s_n/2)^2 + \left[1 - \sqrt{1 - (s_n/2)^2}\right]^2.$$

With a little algebraic manipulation, this formula can be rewritten as

$$s_{2n}^2 = 2 - \sqrt{4 - s_n^2}.$$

Taking the square root of both sides, we finally get

$$s_{2n} = \sqrt{2 - \sqrt{4 - s_n^2}}, \tag{2}$$

which allows us to find s_{2n} in terms of s_n.

Archimedes now implemented the formula by starting with a regular hexagon ($n = 6$), for which each side is equal to the radius 1 (fig. 4.2). Using equation (2) with $s_6 = 1$, he found the length of each side of a regular dodecagon ($n = 12$):

$$s_{12} = \sqrt{2 - \sqrt{4 - 1}} = \sqrt{2 - \sqrt{3}}.$$

Squaring this expression, putting it back in equation (2) and simplifying, he got

$$s_{24} = \sqrt{2 - \sqrt{4 - (2 - \sqrt{3})}} = \sqrt{2 - \sqrt{2 + \sqrt{3}}}.$$

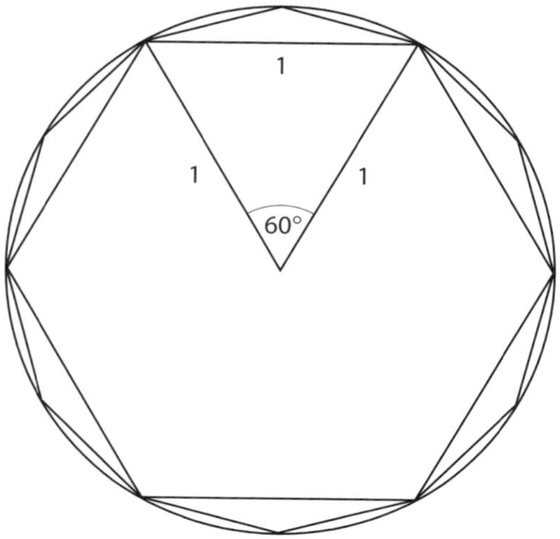

Figure 4.2. A regular hexagon and a regular 12-gon inscribed in a circle

Repeating in this manner, he got

$$s_{48} = \sqrt{2 - \sqrt{2 + \sqrt{2 + \sqrt{3}}}} \, ,$$

and finally

$$s_{96} = \sqrt{2 - \sqrt{2 + \sqrt{2 + \sqrt{2 + \sqrt{3}}}}} \, .$$

To find the perimeter of this 96-sided polygon, we need to multiply s_{96} by 96; and to get the corresponding approximation of π, we need to divide this by 2 (by definition π is one-half the circumference of a unit circle). This gives us the approximation $48\sqrt{2 - \sqrt{2 + \sqrt{2 + \sqrt{2 + \sqrt{3}}}}} \sim 3.14103$, or very nearly $3\frac{10}{71}$.

Archimedes then repeated the process with a series of *circumscribing* polygons of 6, 12, 24, 48, and 96 sides. This calls for the slightly more complicated formula

$$s_{2n} = \frac{2\sqrt{4 + s_n^2} - 4}{s_n}, \tag{3}$$

where s_n now denotes the length of one side of a regular circumscribing n-gon. This equation, like equation (2), can be proved using the Pythagorean theorem

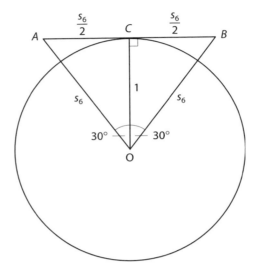

Figure 4.3. A regular hexagon circumscribing a circle (only one side is shown)

(a proof is given in Appendix E). Archimedes again started with a hexagon, this time a circumscribing hexagon (fig. 4.3). To find the length of its side, we note that $\triangle OAB$ is an equilateral triangle, so $OA = OB = AB = s_6$, $AC = s_6/2$, and $OC = 1$. Applying the Pythagorean theorem to the right triangle OAC, we have $OA^2 = OC^2 + AC^2$; that is, $s_6^2 = 1^2 + (s_6/2)^2$, from which we get $s_6 = 2\sqrt{3}/3$.

Archimedes then put this value back into equation (3) to find s_{12}, and then repeated the process three more times to get s_{24}, s_{48}, and finally s_{96}. As before, in order to approximate the value of π from these expressions, we need to multiply each s_n by n and divide by 2. For the 96-sided polygon, Archimedes got the value 3.14271, or very nearly $3\frac{10}{70}$ Since the actual circle is "squeezed" between the inscribed and circumscribing polygons, he concluded that the exact value of π lies between $3\frac{10}{71}$ and $3\frac{10}{70}$.[3]

But Archimedes did more than just approximate π to a remarkable degree of accuracy. He pointed out that the approximation could be improved (at least in principle) by repeating the process again and again, until the desired accuracy was reached. This is because each time the number of sides is doubled, the inscribed and circumscribing polygons squeeze the circle ever more tightly, forcing the value of π to be sandwiched between ever narrower upper and lower limits, like a vise closing on the object between its jaws. This is shown in table 4.1, which gives the values of π to five places for polygons of 3, 6, 12, 24, 48, 96, and 192 sides. The actual value of π, correct to five places, is 3.14159.

TABLE 4.1

n	inscribed n-gon	circumscribing n-gon
3	2.59808	5.19615
6	3.00000	3.46410
12	3.10583	3.21539
24	3.13263	3.15966
48	3.13935	3.14609
96	3.14103	3.14271
192	3.14145	3.14187

It was a brilliant piece of work, all the more so if we consider that the Greeks did not have an efficient method of calculating with numbers; theirs was a hybrid between the old Babylonian base-60 system (see p. 4) and their own system in which each letter of the alphabet had a numerical value (alpha = 1, beta = 2, and so on). It was an *additive* system, sufficient for counting objects but very clumsy when calculations had to be done. Archimedes did it all with papyrus and stylus, or more likely, by drawing figures in the sand.

The idea behind Archimedes' procedure is known as the *method of exhaustion*; it was first formulated by Archimedes' predecessor Eudoxus of Cnidus (see p. 42), but it was Archimedes who made extensive use of it in his work (his finding of the area of a parabolic segment was also based on this method). In this he came tantalizingly close to our modern integral calculus.

Notes and Sources

1. However, doubts have been raised whether Archimedes could have had the technology to polish the reflecting surfaces to the degree necessary to make them effective. See "Briton Questions Archimedes' Feat," *New York Times*, January 10, 1965, and "Recreating an Ancient Death Ray (They Did It with Mirrors)," *New York Times*, October 18, 2005, p. D1.

2. The story of this extraordinary find is the stuff of legend. The manuscript—a copy of a work by Archimedes known as *The Method*—was in the form of a palimpsest, a document that has been written over an earlier text to save on expensive parchment. In this case, the newer script, a twelfth-century religious text, did not completely erase the older writing, making it possible, after painstaking analysis, to read much of the older text. See *The Archimedes Palimpsest* (Christie's auction catalog for October 29, 1998; New York: Christie's, 1998), and *The Works of Archimedes*, ed. Sir Thomas L. Heath (1897; rpt. New York: Dover, 1953; this edition contains all of Archimedes' surviving works, including *The Method*).

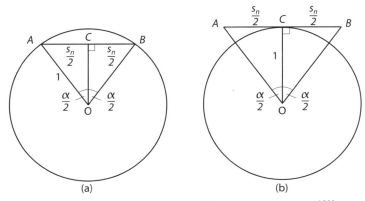

Figure 4.4. Deriving the formulas $s_n = 2\sin\frac{180°}{n}$ (left) and $s_n = 2\tan\frac{180°}{n}$ (right)

3. For $n = 6$, equation (3) gives the simple expression $s_{12} = 4 - 2\sqrt{3}$, but for $n = 24, 48, 96, \ldots$, the expressions for s_{2n} become increasingly cumbersome. For this reason, many authors prefer to derive these formulas using trigonometry. Figure 4.4a shows one side of a regular n-gon inscribed in a unit circle. We have $\alpha = \angle AOB = 360° / n$, so in the right triangle OAC, $\sin\frac{\alpha}{2} = \frac{s_n/2}{1}$ from which we get $s_n = 2\sin\frac{\alpha}{2} = 2\sin\frac{180°}{n}$. Similarly, for the circumscribing n-gon (fig. 4.4b), we have $s_n = 2\tan\frac{180°}{n}$.

These formulas, of course, are much simpler than equations (2) and (3). We should remember, however, that trigonometry did not yet exist in Archimedes' time (it was founded by Hipparchus of Nicaea around 150 BCE), so he could not have availed himself of its methods.

For Archimedes' original derivation (in English translation), see Heath, *The Works of Archimedes*, pp. 91–98.

Translators and Commentators, 500–1500

Unhappy [are] our days, for the study of letters is
dead in our midst, and there is to be found no man
able to record the history of these times.
    —Gregory of Tours (538–594), quoted in David Smith,
    History of Mathematics, vol. 1, p. 183

With Archimedes, the golden age of Greek mathematics reached its zenith.
To be sure, some noted scholars would follow him, but generally the pace and
intensity of Greek mathematics was past its peak. Among his followers, two
names stand out: Apollonius and Diophantus. Apollonius of Perga (ca. 262–ca.
190 BCE) wrote an extensive treatise on the conic sections, a subject not dealt
with by Euclid. In this work he came close to the modern method of coordinates. It was Apollonius who gave the names *ellipse, parabola,* and *hyperbola*
to the curves obtained when a cone is sliced by a plane, depending on whether
the cutting plane is inclined to the base of the cone at an angle smaller than,
equal to, or greater than the angle between the base and the cone's generator.
Only seven of the eight books of this work have survived.[1]

Diophantus of Alexandria, whose exact years of birth and death are unknown (the third century CE is the most probable date) wrote several works,
the most influential by far being his *Arithmetica*, an extensive treatise on number theory and algebraic equations. As with Apollonius, only six of the original thirteen books of this work have survived. In the extant books we find a detailed discussion of the solution to some 130 problems, including equations in
several variables of the first, second, and occasionally higher degrees. Many of
these problems concern various ways of writing sums, differences, or products
of numbers as perfect squares. For example, Proposition 19 of Book III proves
the identity

$$(a^2 + b^2)(c^2 + d^2) = (ac \pm bd)^2 + (ad \mp bc)^2,$$

which shows that the product of two numbers, each of which is a sum of two
squares, is again a sum of two squares, and in fact in two different ways. As an ex-
ample, $65 = 5 \times 13 = (1^2 + 2^2) \times (2^2 + 3^2) = (1 \times 2 \pm 2 \times 3)^2 + (1 \times 3 \mp 2 \times 2)^2,$

giving us the two sums of squares $8^2 + 1^2$ and $4^2 + 7^2$. This identity appeared again in 1202 in Fibonacci's *Liber Abaci*; we will have a chance to use it in Appendix C to construct Pythagorean triples with large components.[2]

The post-Archimedean period also gave rise to several applied mathematicians of note: Eratosthenes of Cyrene (ca. 275–194 BCE), a friend of Archimedes who calculated the circumference of the Earth to a remarkable degree of accuracy (the famous "sieve of Eratosthenes," a method to weed out the prime numbers from the rest of the positive integers, is named after him); Hipparchus of Nicaea (ca. 190–120 BCE), the founder of trigonometry and the author of the first accurate star atlas; and Claudius Ptolemaeus (commonly known as Ptolemy, ca. 85–165 CE), whose great work *Almagest* (in thirteen books, modeled after Euclid's *Elements*), summarized the Greek world picture as it was then known—a geocentric universe in which the Sun, Moon, planets, and stars move around the Earth in perfect circular orbits. These scientists regarded mathematics and astronomy as virtually one subject, a view that would prevail well into the sixteenth century; indeed, many of the pioneers of science in the Renaissance were equally productive in both fields, among them Copernicus, Galileo, and Kepler.

The last noted name in the long line of Greek mathematicians is Pappus of Alexandria, who most likely lived in the third century CE. He wrote commentaries on several of Euclid's works, including the *Elements*, but nearly all of them are lost, and we know about them only through later writers. The one work that did partially survive is his *Mathematical Collections*, in eight books, of which only the last six have come to us intact. These books contain treatises on proportion, on solids and spheres, and on various planar curves. Book V discusses the *isoperimetric problem*—to find the planar figure of greatest area enclosed in a given perimeter (this subject is nowadays studied under the calculus of variations, an extension of ordinary calculus concerned with maximizing or minimizing definite integrals of functions, rather than the functions themselves). Also in the *Collections* are two theorems discovered by Pappus himself: one, in Book VII, on finding the surface area and volume of solids of revolution (this theorem is now known as Guldin's theorem, after the Swiss Paul Guldin [1577–1643], although the latter, in all likelihood, was aware that Pappus had preceded him by a thousand years);[3] the other, in Book IV, is an extension of the Pythagorean theorem:

> Let *ABC* be any triangle, and let *ABDE* and *ACFG* be two
> parallelograms built on the sides *AB* and *AC*, respectively (fig. 5.1).
> Extend *DE* and *FG* until they meet at *H*. Draw *BM* = *CN*, each
> parallel to *HA* and equal to it. This produces the parallelogram
> *BMNC*. Pappus's theorem says that the area of this parallelogram is
> equal to the sum of the areas of the original parallelograms.

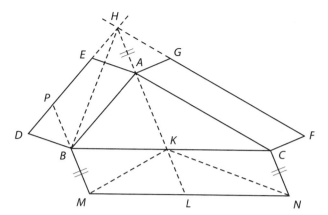

Figure 5.1. Pappus's theorem

The proof is along lines similar to that of the Pythagorean theorem in Euclid I 47. Extend *HA* until it intersects *BC* at *K* and *MN* at *L*. This divides *BMNC* into two parallelograms, *BKLM* and *CKLN*. We claim that the area of *BKLM* is equal to that of *ABDE*, and similarly the area of *CKLN* is equal to that of *ACFG*.

To show this, extend *MB* until it meets *DE* at *P*. We have $\mathbf{A}_{ABDE} = \mathbf{A}_{ABPH}$ (where **A** denotes area), the two parallelograms having a common base *AB* and opposite sides *DE = PH* lying on the same line. Now draw the diagonal *HB*, dividing *ABPH* into two congruent triangles *ABH* and *PBH*. We have $\mathbf{A}_{ABPH} = 2\mathbf{A}_{ABH}$. But triangles *ABH* and *BKM* have the same area, having equal bases *HA* and *BM* and vertices *B* and *K* that lie on parallel lines. Thus,

$$\mathbf{A}_{ABDE} = 2\mathbf{A}_{BKM} = \mathbf{A}_{BKLM}.$$

In exactly the same manner, we have

$$\mathbf{A}_{ACFG} = 2\mathbf{A}_{CKN} = \mathbf{A}_{CKLN}.$$

Adding the two equations, we finally get

$$\mathbf{A}_{ABDE} + \mathbf{A}_{ACFG} = \mathbf{A}_{BMNC},$$

which is Pappus's theorem. The Pythagorean theorem follows as a special case when angle *A* is a right angle and the two parallelograms are squares.

The *Collections* contains many more beautiful and innovative results, the product of a creative mind that was destined to close the golden age of Greek geometry. With Pappus, the thousand-year period of Greek eminence in mathematics, begun with Thales around 600 BCE, came to an end.[4]

❖　❖　❖

The decline of Greek mathematics—and Greek culture in general—did not happen in a vacuum. Great political and social changes were in the air. In 212

BCE Syracuse was taken by the Romans (as already noted, it is thanks to this event that we know the exact year of Archimedes' death). Soon to follow was Carthage, and within a hundred years the entire Mediterranean basin became part of the Roman Empire. Egypt, under the rule of the Ptolemy dynasty, kept its relative independence until 30 BCE, by which time it too fell under Roman domination. The Roman administration generally did not interfere with the cultural and commercial life of the nations under its rule, so long as they paid their taxes and did not revolt against their rulers. But this relatively peaceful coexistence would soon change. In 330 CE Emperor Constantine I embraced Christianity and declared the city of Byzantium his capital, renaming it Constantinople. Sixty years later, in 389, a cultural crime of unimaginable consequences took place, the burning of the famed library of Alexandria during a riot of Christians against pagans. Up in flames went its prized collection of over half a million scrolls—nearly the entire literary and scientific heritage of the ancient world.

Shortly thereafter, in 392, Emperor Theodosius declared Christianity the official religion of the Roman Empire. On his death in 395, the empire was split into the eastern and western parts. Although the Hellenistic world kept its identity as part of the eastern dominion, its role as a cultural and intellectual powerhouse quickly declined. In 476 the city of Rome fell to the Vandals, marking the end of the once mighty Roman Empire. In 529 Emperor Justinian closed the Academy in Athens, founded by Plato nearly nine hundred years earlier. Its few remaining scholars fled to Egypt and Persia, where scattered centers of learning continued to function despite adverse conditions. The Alexandrian library was partially rebuilt, though it never regained its original glory. Then, in 641, the Arabs conquered Alexandria. Under the orders of its caliph, the remnants of the library were burned again, and the role of Alexandria as the ancient world's foremost center of study came to an end. The Dark Ages were about to begin.[5]

Still, a few scholars kept the tradition of learning alive by writing commentaries on the surviving works of their predecessors; among them, two names stand out: Theon and Proclus.

Theon of Alexandria lived at the end of the fourth century CE. Around 390 he wrote a revised version of the *Elements*, and it was this revision that would be the basis of most of the later editions of Euclid. He also wrote an eleven-book commentary on Ptolemy's *Almagest*. Theon's name will forever be associated with his daughter Hypatia (ca. 370–415), who studied under him and later became a mathematician in her own right. She is said to have written commentaries on the works of Diophantus and Apollonius. According to tradition, her reputation was so great that she was called upon to head the Neoplatonic school in Alexandria. Hypatia was smart, articulate, and beautiful. This

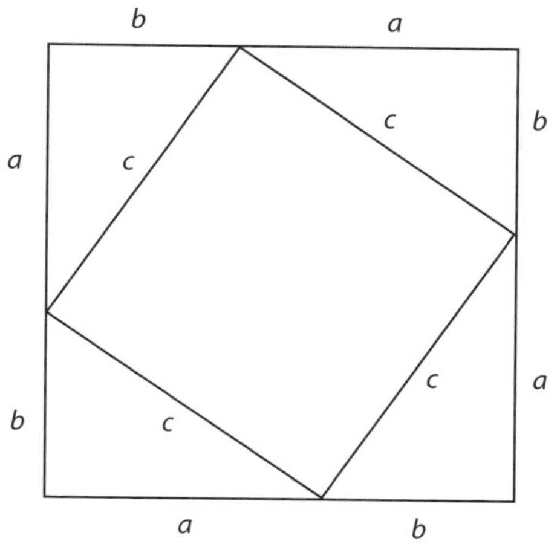

Figure 5.2. Dissection of a square

brought upon her the ire of an incensed religious mob; she was accused of preaching paganism and was cruelly murdered, ending the life of the first woman mathematician in history.[6]

Proclus (ca. 412–485 CE) was born in Byzantium, studied in Alexandria, and later became head of the Athenian Academy. He is chiefly known for his *Eudemian Summary*, a work that includes his own commentary on Book I of the *Elements* and a historical outline of Greek geometry up until Euclid's time; this work was based on fragments of an earlier work, *History of Geometry* (in four books) by Eudemus, a student of Aristotle. In the *Eudemian Summary* we find the famous saying attributed to Euclid, "There is no royal way to geometry." It is from Proclus's commentary that we may infer a possible proof of the Pythagorean theorem by Pythagoras, namely, a proof by *dissection*:

> Consider a square of side $a + b$ (fig. 5.2). Connect the dividing points between segments a and b of each side to form a tilted square, and call its side c. The original square is thus dissected into five parts—four congruent right triangles of sides a and b and hypotenuse c, and an inner square of side c. A different dissection is shown in figure 5.3. Comparing the areas of the two figures, we have

$$4\frac{ab}{2} + c^2 = 4\frac{ab}{2} + a^2 + b^2,$$

from which we get $c^2 = a^2 + b^2$. This is essentially the "Chinese proof" already referred to on page 25, and to which we will come back shortly. The proof involves some subtleties, such as demonstrating that the four right triangles in the one figure are indeed congruent

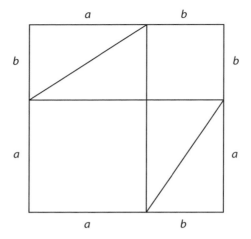

Figure 5.3. A different dissection

to those in the other. This, in turn, is based on the theorem that the sum of the angles in a triangle is equal to two right angles; the *Eudemian Summary* indicates that this last theorem was already known to the Pythagoreans.[7]

❖   ❖   ❖

Living in near isolation from the West, the vast kingdom of China had already developed into a highly advanced civilization, whose achievements included the invention of paper, block printing, gunpowder, and the magnetic compass—this hundreds of years before these inventions became known to Europeans. The long Chinese history, going back at least to the third millennium BCE, boasted writers on poetry, philosophy, astronomy, and mathematics, as well as on practical subjects such as agriculture and engineering. And the Chinese had their own burning of libraries: in the year 213 BCE the emperor Shi Huang-ti ordered the destruction of all existing books, a decree that obliterated much of the cultural heritage of previous generations of writers; luckily, some books escaped the flames, while others were later reconstructed from memory.

One of the oldest Chinese mathematical works known to us is the *Chao Pei Suan Ching* ("The arithmetical classic of the gnomon and the circular paths of heaven"), the date of which is uncertain but is most likely from the time of the Han dynasty (206 BCE–221 CE), and possibly even earlier.[8] Though chiefly a work on the calendar, it has some material on earlier Chinese mathematics. The first part is in the form of a dialogue between the ruler Chou Kung and a person by the name of Shang Kao, the subject being the properties of right triangles. Here we find the verse, "Break the line and make the breadth 3, the length 4; then the distance between the corners is 5," clearly a reference to the

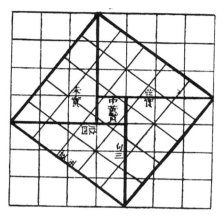

Figure 5.4. Chinese demonstration of the Pythagorean theorem

3-4-5-sided triangle.[9] The Pythagorean theorem is stated in words: "Multiply both the height of the post [the gnomon] and the shadow length [the base] by their own values, add the squares, and take the square root." That is, $c = \sqrt{a^2 + b^2}$. There is a diagram demonstrating the (3, 4, 5) triangle (fig. 5.4), but it can easily be generalized to any right triangle. The diagram is accompanied by an explanatory note:

> Thus, let us cut a rectangle (diagonally), and make the width (*kou*) 3 (units) wide, and the length (*ku*) 4 (units) long. The diagonal (*ching*) between the (two) corners will then be 5 (units) long. Now after drawing a square on this diagonal, circumscribe it by half-rectangles like that which has been left outside, so as to form a (square) plate. Thus the (four) outer half-rectangles of width 3, length 4, and diagonal 5, together make (*te chheng*) two rectangles (of area 24); then (when this is subtracted from the square plate of area 49) the remainder (*chang*) is of area 25. This (process) is called "piling up the rectangles" (*chi chu*).[10]

Here, "rectangle" refers to any of the four 3 by 4 corner rectangles, and the "half-rectangles" are the 3-4-5–sided right triangles shown in the figure. To "circumscribe it [the square built on the diagonal] by half-rectangles" simply means to surround this tilted 5 by 5 square with four 3-4-5–sided right triangles identical to the one we started with. This creates an outer 7 by 7 square ("square plate of area 49"). Subtracting from this square the combined area of the four triangles, we get $49 - 24 = 25$, the area of the tilted square.

Of course, this would not be considered a proper proof in the Greek style, in which a sequence of logical deductions follows in strict order, starting from a small number of agreed-upon, self-evident axioms. The Chinese idea of a proof was to produce a convincing visual demonstration from which the general case could be inferred. To quote the eminent scholar Joseph Needham, "In the Chinese approach, geometrical figures acted as a means of transmutation whereby numerical relations were generalised into algebraic form."[11] The algebraic form in this case would be the identity $c^2 = (a + b)^2 - 4\frac{ab}{2} = a^2 + b^2$, where a and b are the length and width of the corner rectangles.[12]

This kind of proof fitted nicely with an old Chinese tradition of dissecting the parts of a planar shape and recombining them in a different way, as in the familiar tangram puzzle. Indeed, the phrase "piling up the rectangles" in the quotation above refers to dissecting the square in figure 5.4 and then recombining it until the equality of areas becomes obvious. To enhance the visual effect, the diagram in later editions was often color-coded, the small inner square being yellow and the rectangles around it red.[13] A thousand years later, an identical proof was given by the Hindu mathematician Bhaskara ("the Learned," 1114–ca. 1185); he simply drew the inner (tilted) square in figure 5.4 without any comments save for the word "see!"—the kind of "proof without words" that became popular in modern mathematics journals.

Because the Chinese words for the width and length of a rectangle are *kou* and *ku*, respectively, the Pythagorean theorem has been known in China as the *kou-ku* theorem. As such it appears in numerous problems in Chinese writings, mostly of a practical nature. We bring here one example, the *Broken Bamboo*. It first appeared in the *Chiu Chang Suan Shu* ("Nine chapters on the mathematical art"), probably dating back to the Han dynasty and described by Needham as "perhaps the most important of all Chinese mathematical books."[14] This problem is often accompanied by an illustration (fig. 5.5) that first appeared in a work by Yang Hui entitled *Hsiang Chieh Chiu Chang Suan Fa*, dating to 1261 CE. The problem reads:

> There is a bamboo 10 chih high, the upper end of which, being broken, touches the ground 3 chih from the foot of the stem. What is the height of the break?[15]

Let the breaking point be x chih above the ground, and let the height from there to the top be a (fig. 5.6). Denoting the horizontal distance from the tip of the broken part to the foot of the bamboo by b, we have

$$b^2 = a^2 - x^2 = (a + x)(a - x).\tag{1}$$

But $a + x = h$ is the total height of the bamboo, so we can write equation (1) as

$$b^2 = h\,[(h - x) - x] = h(h - 2x).$$

Solving this last equation for x, we get $x = (h^2 - b^2)/2h$. This solution is given

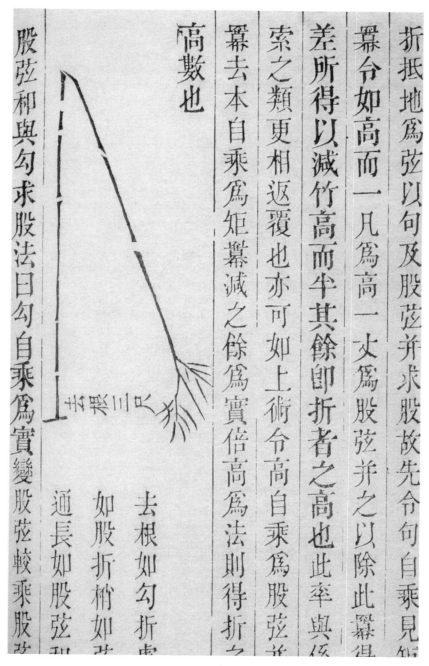

折抵地爲弦以句及股弦并求股故先令句自乘見矩

冪令如高而一凡爲高一丈爲股弦并之以除此冪得

差所得以減竹高而半其餘即折者之高也此率與係

索之類更相返覆也亦可如上術令高自乘爲股弦并

冪去本自乘爲矩冪減之餘爲實倍高爲法則得折之

高數也

股弦和與勾求股法曰勾自乘爲實變股弦較乘股弦

去根如勾折處

如股折枝如莖

通長如股弦和

Figure 5.5. The Broken Bamboo

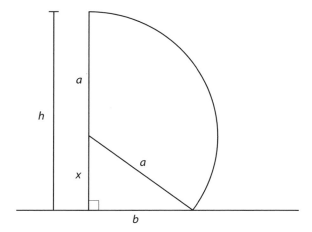

Figure 5.6. The Broken Bamboo: a schematic diagram

by Yang Hui verbally as "do such and such" instructions. Putting $h = 10$ and $b = 3$ in the solution, we get the answer $x = 91/20$ chih.

To conclude our discussion of the Chinese role in the Pythagorean theorem, I will quote from George Gheverghese Joseph's book *The Crest of the Peacock: Non-European Roots of Mathematics*:

> The importance of the *kou ku* theorem in establishing algebraic geometry and its contribution to the broader development of Chinese algebra cannot be overestimated. It founded a tradition in geometric reasoning which belies the notion that all mathematical traditions not influenced by the Greeks were essentially algebraic and empirical.[16]

❖ ❖ ❖

In contrast to China, the Indian subcontinent was at the crossroads of several civilizations. To the north lay Tibet and Afghanistan, accessible only through high passes in the Himalayas. To the northwest were Persia and the great plains of Central Asia, and to the west, the Arabian Peninsula and the Mediterranean beyond. Moreover, India was easily accessible by sea, being flanked on the east by the Bay of Bengal and on the west by the Arabian Sea. Consequently, Indian culture has been greatly influenced by the multitude of civilizations across its borders, and in turn it exerted its influence on them. And this included mathematics.

The earliest Hindu writings on mathematics were an offspring of Hindu religious practices. A group of writings known collectively as the *sulbasturas* deals with the dimensions of sacrificial altars (*vedi*), a subject of great importance in Hindu religion. One of these *sulbasturas*, by an author named Baudhayana, dates back to perhaps 600 BCE or earlier, to the time of Thales.[17] In

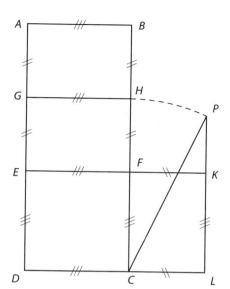

Figure 5.7. Squaring a rectangle

this work we find the statement, "The rope which is stretched across the diagonal of a square produces an area double the size of the original square"—in other words, the special case of the Pythagorean theorem for the 45-45-90–degree triangle. A later *sulbastura*, by an author named Katyayana, states the general theorem: "The rope [stretched along the length] of the diagonal of a rectangle makes an [area] which the vertical and horizontal sides make together." The text then gives instructions on how to build a trapezoid-shaped altar of length 36 *pados* (literally, feet); this involves various auxiliary lines that form the Pythagorean triangles (5, 12, 13), (8, 15, 17), (12, 16, 20), (12, 35, 37), (15, 20, 25), and (15, 36, 39).[18]

The *sulbastura* of Baudhayana mentioned earlier also has instructions on how to "square a rectangle"—how to construct a square equal in area to a given rectangle. Let the rectangle be *ABCD* (fig. 5.7). Draw *EF* equal to *AB* and perpendicular to *BC*, forming the square *EFCD*. Draw *GH* midway between *AB* and *EF* and parallel to either. Now rotate rectangle *ABHG* by 90 degrees and move it to the position *FKLC*, so that *KL = AB* and *LC = BH*. With center at *C*, swing an arc of radius *CH*, cutting the extended line *LK* at *P*. We have $LP^2 = CP^2 - CL^2 = CH^2 - CL^2 = (CH + CL) \times (CH - CL) = (CH + HB) \times (CH - HB) = (CH + HB) \times (CH - HF) = CB \times CF = CB \times CD$ (note that all line segments here are nondirected, so that *CL = LC*, etc.). Thus the square with side *LP* is the required square.[19]

Also in the *sulbasturas* are instructions on how to construct a square whose area is the sum of two given squares. Let *ABCD* and *EFGH* be the given squares, with *AB > EF* (fig. 5.8). Along *AB* allocate a segment *AP = EF*, and

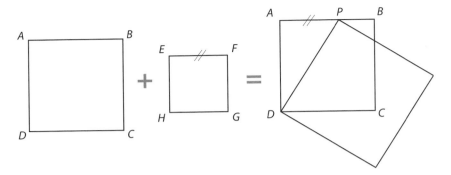

Figure 5.8. Constructing a square whose area is the sum of two squares

connect D with P. We have $PD^2 = AP^2 + AD^2 = EF^2 + AD^2$, so that the square built on PD is the required square.

These examples clearly show that the Hindus had mastered the Pythagorean theorem at least as early as Pythagoras, and that they knew how to apply it to a whole range of practical problems. But as to the demonstration they used to prove the theorem—if indeed they proved it—we are again in complete darkness. As with the Chinese, we may speculate that they used some kind of a dissection proof, but it is just that—a speculation.[20]

❖   ❖   ❖

We now return to the Mediterranean basin and the Middle East. After the death of the Prophet Mohammed in 632 CE, Islam spread northward and westward with lightning speed. Before the century was out, the Islamic Empire would stretch from Persia in the east to the Atlantic Ocean in the west and from Central Asia in the north to the Sahara Desert in the south. This enormous expansion reached its peak in 711 when the Moslems entered Spain and established a caliphate that would last over eight hundred years.

The new rulers of this vast empire were fierce warriors, but they also showed a great passion for learning. Centers of study were established in Baghdad, in Samarkand (in modern Uzbekistan), and in Cordoba in Spain, and scholars of various ethnicities and religions—Jewish, Moslem, and Persian—were encouraged to settle there. They studied, interpreted, and evaluated what was left of the ancient Greek writings—those books that survived the demise of the Alexandrian library and had been kept in the relative safety of monasteries in Constantinople, Damascus, and Jerusalem. Even more important to history, these scholars translated numerous Assyrian, Greek, and Sanskrit works into Arabic, from which they would be retranslated into Latin and thus made known to the West. Thanks to these scholars, the flame of learning was kept alive while Europe plunged into the Dark Ages.

In the year 762 the caliph al-Mansur (712?–775) moved his capital to Baghdad, rebuilding it and making it a major center of learning, "a second Alexandria." Under his successor Harun al-Rashid (Aaron "the Just," reigned 786–809), a major translation project of Euclid's *Elements* and Ptolemy's *Almagest* was begun, a task that was completed under his son al-Mamun (reigned 809–833).[21] Al-Mamun built an observatory in Baghdad from which he supervised a variety of geodetic surveys. In his court lived perhaps the greatest of all Arab mathematicians, Mohammed ibn Musa al-Khowarizmi (born in Khwarezm, east of the Caspian Sea, ca. 780 and died ca. 850), whose classic work *'ilm al-jabr wa'l muqabalah* ("the science of reduction and cancellation") is considered the first important work on algebra (indeed, the word "algebra" comes from the *al-jabr* in the title). A second work by him, existing only in Latin translation, was *Algoritmi de numero Indorum*, in which he advocated the use of the Hindu base-ten numeration system (the modern word "algorithm" is a corruption of the name al-Khowarizmi). These two works would later have a profound influence on the advance of mathematical literacy in the West.

Of special interest to us here is Tabit ibn Qorra ibn Mervan, Abu-Hasan, al-Harrani (826–901); as indicated by his name, he was a native of Harran in Mesopotamia. He was a physician, philosopher, and mathematician, in which role he was a pioneer in applying algebraic methods to geometry. Ibn Qorra revised an earlier translation of the *Elements* (fig. 5.9) and translated several other Greek works written between the times of Euclid and Ptolemy. He worked on spherical trigonometry—a subject closely related to astronomy— and on parabolas and paraboloids. His son and two of his grandsons continued in his footpath as mathematicians and translators.

Ibn Qorra is of interest to us because of his generalization of the Pythagorean theorem. Let *ABC* be any triangle (fig. 5.10). From the top vertex *A* draw lines *AM* and *AN* such that $\angle AMB = \angle ANC =$ angle *A*. Triangles *ABC*, *MBA*, and *NAC* are therefore similar, each having one angle in common with the original triangle and a second angle equal to *A*. Therefore,

$$AB/BC = MB/AB, \quad \text{from which} \quad AB^2 = BC \times MB,$$

and

$$AC/BC = NC/AC, \quad \text{from which} \quad AC^2 = BC \times NC.$$

Adding the two equations, we have

$$AB^2 + AC^2 = BC \times (MB + NC),$$

which is ibn Qorra's theorem. The Pythagorean theorem follows as a special case when *A* is a right angle, in which case points *M* and *N* coincide and $MB + NC = BC$ (as before, all line segments are nondirected).[22]

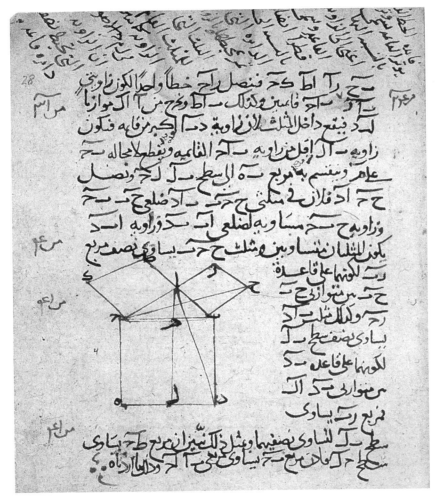

Figure 5.9. Ibn Qorra's translation of Euclid, showing the Pythagorean theorem

Due to their geographic location between India and China in the east and the Mediterranean nations on the west, the Moslems were in a unique position to transmit what was left of the ancient literary and scientific heritage to Europe. It is one of the ironies of history that the golden age of Islamic science roughly coincided with the Dark Ages in Europe, providing some sense of continuity between the old world and the new. But political and social upheavals soon put an end to the golden era. In 1258 Baghdad fell to the Mongols, and its cultural institutions were destroyed. For a brief period, a new center of learning was established in Samarkand under the enlightened rule of

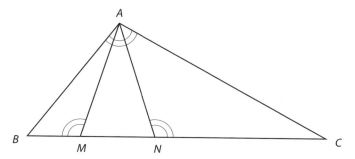

Figure 5.10. Ibn Qorra's theorem

Ulugh Beg (1393–1449); it boasted the largest observatory in the pretelescopic era, whose remnants still exist today. Ulugh Beg's assistant Jemshid ibn Mes'ud ibn Mahmud, Giyat ed-din al-Kashi (d. 1429 or 1436), wrote several works on arithmetic and geometry. He calculated π to an unprecedented accuracy of sixteen decimal places, using Archimedes' method of exhaustion (see p. 55) with inscribed and circumscribing polygons of $3 \times 2^{28} = 805,306,368$ sides! But by the time of his death, the hegemony of the eastern part of the Islamic Empire has by and large ended.

❖   ❖   ❖

As the centers of learning in the east declined, new centers were being established at the western end of the caliphate, in Spain. Christian, Jewish, and Moslem scholars were engaged in translating the classic Greek works from Arabic into Latin and occasionally Hebrew, often adding their own commentaries. One of the most prominent of these translators was Gherardo of Cremona (also known as Gerard or Gerardus, 1114–1187), who was likely born in Lombardy (Italy), although some claim him as an Andalusian. His translation of Euclid's *Elements* and Ptolemy's *Almagest* made these works accessible to European scholars. He seems to have been the first to use the word *sinus* for the half-chord spanned by a central angle in a circle, a precursor of our modern sine function.

Slowly, Europe was awakening from its long slumber. The first European university was founded in Bologna in 1088, to be followed by those of Paris (1200), Oxford (1214), Padua (1222), and Cambridge (1231). The centers of learning were gradually moving away from the Catholic Church to more secularly oriented institutions, although it would take another four centuries before this shift was completed. There were numerous setbacks: the Hundred Year War between England and France (1338–1453) sapped the energy of much of Europe, to be followed by the Black Death, an epidemic of biblical proportions that decimated well over one-third of Europe's population.

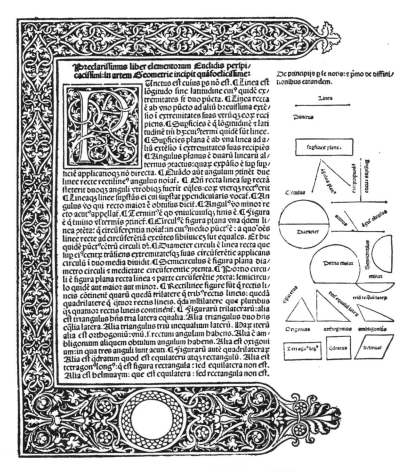

Figure 5.11. The first printed edition of Euclid (Venice, 1482)

In 1453 Constantinople fell to the Turks, who changed its name to Istanbul. This event is traditionally regarded as the end of the Middle Ages, though not of the Dark Ages: the expulsion of the Jews from Spain in 1492 deprived the country of much of its commercial, artistic, and intellectual elite, a loss from which it has never completely recovered. That same year saw the end of the Moorish presence in Spain; their last stronghold, the magnificent palace of the Alhambra in Granada, would one day be the inspiration for a young artist named M. C. Escher. And as if these events were not enough for a single year, it was in 1492 that Christopher Columbus landed in the New World, opening European trade to vast new lands and unimaginable riches. The New Age had begun with a bang.

❖ ❖ ❖

98

Eglie una torre che e
alta 40 braccia z dap
pie uipaiſa uno fiume
che e largbo 30 brac
cia. uo ſapere quanto
ſara lungba una fune
che ſia appicata alla ri
ua del fiume z alla ci
ma della torre

40————30
40· 30

1600
900

laradice di 500
ſara lunga 50 brac
cia

Eglie unalbero inſu la
riua dun fiume elqua
le e alto 50 braccia el
fiume e largbo 30 bra
cia z per fortuna di ué
to ſiruppe intal luogo
che lacima dellalbero
toccaua lariua del fiu
me. Uo ſapere quante
braccia ſene ruppe z
quanto nerimaſe ritto

50————30
50 500 30
100| 900
1600

rimaſe ritto 16 brac
cia z 34 braccia ſene
ruppe

Figure 5.12. A page from Filippo Calandri's book on arithmetic, showing two problems involving the Pythagorean theorem

But an intellectual "bang" of an even greater significance had already taken place some years earlier. In 1454 Johannes Gutenberg, a German living in Mainz, invented the movable-type printing press, launching it with a special, three-hundred-copy print run of the Holy Bible in an exquisitely decorated edition. Soon, books of every kind were printed in mass quantities and sold all over Europe; by 1500, some nine million copies of 30,000 different titles were available. The laborious, age-old art of copying and recopying manuscripts by hand was gone for good.

It was not long before the classic works of mathematics appeared in print. The first printed edition of the *Elements* was published in Venice in 1482; it was a work of art as much as of science, with many color illustrations accompanying the text (fig. 5.11). Nine years later there appeared in Florence a book on arithmetic by Filippo Calandri that contained the first illustrated "word problems" published in Italy (fig. 5.12 shows two problems involving the Pythagorean theorem). But it was not until 1570 that the first English edition of the *Elements* was published. The translation has been attributed to Sir Henry Billingsley, who in 1596 became sheriff and Lord Mayor of London; a preface was written by John Dee (1527–1608), one of the original fellows of Trinity College, where Newton would be a professor a century later. Still, with a few notable exceptions (for example, Descartes's *La Géométrie*, his 1637 treatise on analytic geometry, which was written in French), Latin continued to be the international language of science and letters for another hundred years. It was only toward the end of the seventeenth century that scientists began writing original texts in their native languages, thereby making mathematics accessible to the general public in each country. No longer the exclusive domain of scholars, it was now available to all who wanted to open their minds to its teachings.

Notes and Sources

1. For a fuller survey of Apollonius's work, see Sir Thomas L. Heath, *A Manual of Greek Mathematics* (Oxford, U.K.: Oxford University Press, 1931), pp. 352–376. See also Eves, pp. 171–175 and 191–192.

2. For a fuller survey of Diophantus's work, see Heath, *Greek Mathematics*, chap. 17. See also Eves, pp. 180–182 and 197.

3. Smith, vol. 1, pp. 433–434. Smith goes so far as to call Guldin a plagiarist for including Pappus's theorem in his own writing without crediting its true discoverer.

4. For a fuller survey of Pappus's work, see Heath, *Greek Mathematics*, chap. 16. See also Eves, pp. 182–184 and 197–199.

5. This brief historical sketch is based on Eves, p. 164.

6. For her biography, see Marilyn Bailey Oglivie, *Women in Science—Antiquity through the Nineteenth Century: A Biographical Dictionary with Annotated Bibliography* (Cambridge, Mass.: Massachusetts Institute of Technology Press, 1986), pp. 104–105.

7. Eves, pp. 80–81.

8. I have used here the spelling and translation of the title as given in Joseph Needham's exhaustive work, *Science and Civilisation in China* (Cambridge, U.K.: Cambridge University Press, 1959, in 10 vols.), vol. 3, p. 19; other sources give different versions. According to Needham, the word *Chao* may refer to the Chao dynasty, but it could also mean "circumference," an allusion to the Sun's circular orbit around the Earth. "Pei" refers to a gnomon, a rod with which one can measure the Sun's position by the length of the shadow it casts, as in a sundial. For a detailed background on the *Chao Pei Suan Ching*, see ibid., pp. 19–24.

9. Smith, vol. 1, p. 31.

10. Needham, *Science and Civilization*, vol. 3, pp. 22–23. The parentheses are Needham's.

11. Ibid., p. 24.

12. Equivalently, we could compare the area of the tilted square in figure 5.4 with that of the small inner square, leading to the equation $c^2 = (a - b)^2 + 4\frac{ab}{2} = a^2 + b^2$.

13. Needham, *Science and Civilization*, p. 96.

14. Ibid., p. 25.

15. I have used the translation given in George Gheverghese Joseph, *The Crest of the Peacock: Non-European Roots of Mathematics* (1991; rpt. Princeton, N.J.: Princeton University Press, 2000), p. 186.

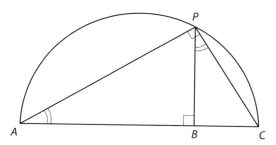

Figure 5.13. Squaring a rectangle: Euclid's construction

16. Ibid., p. 187. Much more on the Pythagorean theorem in China, including numerous problems based on it, can be found in Frank J. Swetz and T. I. Kao, *Was Pythagoras Chinese? An Examination of Right Triangle Theory in Ancient China* (University Park, Penn.: Pennsylvania State University Press, and Reston, Va.: National Council of Teachers of Mathematics, 1977). This book also has an extensive bibliography.

17. It must be borne in mind that all dates of early Hindu writings are uncertain. These writings have survived only because they have been copied over and over, the date of the copy often being mistakenly taken as the date of the original.

18. The preceding paragraph and the following two are based on Joseph, *Crest of the Peacock* pp. 224–230. The two formulations of the Pythagorean theorem are quoted verbatim from there. Joseph gives an excellent survey of Hindu mathematics from 800 BCE to modern times, a subject only marginally covered in most books on the history of mathematics.

19. One cannot fail to note how much more involved this construction is than the one given by Euclid in Proposition 14 of Book II and based entirely on similarity: Let the rectangle be *ABCD*. Draw *AB* and *BC* on the same line, meeting end-to-end at *B* (fig. 5.13). Draw a semicircle with diameter *AC*. At *B* erect a perpendicular to *AC*, meeting the semicircle at *P*. We have $\angle BAP = \angle BPC$ and also $\angle APC = 90°$. Therefore, triangles *ABP* and *PBC* are similar, and so $AB/BP = PB/BC$, from which $BP^2 = AB \times BC$. Thus *BP* is the side of the required square.

20. Smith (vol. 1, p. 97) goes even further: "There is no reason for believing that the Hindus had the slightest idea of the nature of a geometric proof." One may assume that by "geometric proof" he meant a proof in the Greek style—a sequence of logical deductions that follow from a small set of axioms. But as we have seen, the Chinese and Hindu concept of a proof was quite different.

21. The English spelling of Arabic names varies greatly from one author to another. To maintain some degree of consistency, I have followed here the spelling in Smith.

22. See Robert Shloming, "Thâbit ibn Qurra and the Pythagorean Theorem," in *From Five Fingers to Infinity: A Journey through the History of Mathematics*, ed. Frank J. Swetz. Chicago and La Salle, Ill.: Open Court, 1995), chap. 43.

François Viète Makes History

The Renaissance was the golden age of the
amateur mathematician.
—Petr Beckmann, *A History of π*, p. 97

Nearly 1800 years after Archimedes, a French lawyer and freelance mathematician by the name François Viète (1540–1603) made history by appending a certain algebraic formula with the word "etcetera," implying that the procedure called for by the formula could be repeated again and again, ad infinitum. The Greeks, of course, had been keenly aware of the existence of infinity, but lacking the algebraic tools to deal with it, they shunned it from their mathematical world. It remained a controversial subject for the next two millennia, to be avoided at all cost. Viète's seemingly benign etcetera at once broke this age-old taboo, and with lightning speed infinity assumed central stage in mathematics.

Viète (or Vieta, the Latinized version of his name) was one of the first in a long line of French mathematicians who also served their country in the civil services, in politics, and in the military. Although he practiced mathematics merely as a pastime, his reputation was such that King Henry IV called upon him to decipher the secret code the Spanish army was using in its war against France. So successful was Viète that the Spaniards, amazed that their code had been broken, accused the French of using sorcery, "contrary to the practice of the Christian faith."[1]

More important to the future of mathematics, however, was Viète's introduction of symbols into algebra; up until his time, algebraic statements were expressed verbally, making it extremely difficult to perform routine algebraic operations. Viète devised a system in which known quantities were denoted by consonants and unknowns by vowels. This system would later be changed by Descartes to the form we know today (a, b, c, for constants, x, y, z for variables), but it was Viète who initiated the transition from verbal to symbolic algebra, a transition considered to be one of the most important developments in the history of mathematics.

Viète also made significant contributions to trigonometry. He showed how to use algebraic methods to solve trigonometric equations and vice versa, and he was the first to express $\sin nx$ and $\cos nx$ in terms of $\sin x$ and $\cos x$ (he did so for all integer values of n up to 10; the general case was found by Jakob Bernoulli in 1702, over a hundred years after Viète's work). Essentially, Viète transformed trigonometry from a subject limited to the solution of triangles into the analytic subject it has become since.

Toward the end of his life, Viète embroiled himself in several controversies that tarnished his reputation. He was involved in a bitter dispute with the German mathematician Christopher Clavius (1537–1612) over the reformation of the Julian calendar, ordered by Pope Gregory XIII in 1582. Viète's vitriolic attacks on Clavius, who was the pope's adviser on the matter, earned him many enemies. Viète was also opposed to the Copernican system that put the Sun, instead of the Earth, at the center of the universe. He was at once a great innovator and a conservative shackled by old conventions, a product of the transition from the old world to the new.[2]

❖ ❖ ❖

Viète's groundbreaking formula, which he discovered in 1593, expressed the number $2/\pi$ as an *infinite product*:

$$\frac{2}{\pi} = \sqrt{\frac{1}{2}} \cdot \sqrt{\frac{1}{2} + \frac{1}{2}\sqrt{\frac{1}{2}}} \cdot \sqrt{\frac{1}{2} + \frac{1}{2}\sqrt{\frac{1}{2} + \frac{1}{2}\sqrt{\frac{1}{2}}}} \cdots$$

(it is often written in the equivalent form

$$\frac{2}{\pi} = \frac{\sqrt{2}}{2} \cdot \frac{\sqrt{2+\sqrt{2}}}{2} \cdot \frac{\sqrt{2+\sqrt{2+\sqrt{2}}}}{2} \cdots;$$

as already mentioned, Viète used "etc." instead of the three dots). It shows that the value of π can be computed, at least in principle, by repeatedly applying the arithmetic operations of addition, multiplication, division, and square-root extraction, all on the number 2. We should note, however, that due to its slow rate of convergence, the *practical* value of Viète's product for calculating π is rather limited.

Viète derived his product essentially as Archimedes had done, by inscribing regular polygons of ever more sides in a unit circle. But he departed from his predecessor in two ways: instead of considering perimeters, Viète found the *areas* of the inscribed polygons; and he started with a square instead of a hexagon. Before we show his derivation, we need to establish a trigonometric identity, the *half-angle formula* for the cosine function. In doing so, we will be reminded of the central role the Pythagorean theorem plays in trigonometry.

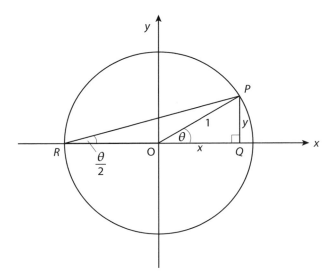

Figure 6.1. Deriving the half-angle formula

Figure 6.1 shows a unit circle with its center at the origin of a rectangular coordinate system. Let P be a point on this circle, and let the radius vector OP form an angle θ with the positive x-axis. In triangle OPQ we have $\cos \theta = x$, $\sin \theta = y$. By a well-known theorem (*Euclid* III, 20), the circumferential angle QRP is equal to one-half the central angle subtending the same arc; that is, $\angle QRP = \frac{1}{2} \angle QOP = \theta/2$. In the right triangle RPQ we therefore have

$$\cos \theta/2 = \frac{adjacent\ side}{hypotenuse} = \frac{RQ}{RP}$$
$$= \frac{RO + OQ}{RP} = \frac{1 + x}{\sqrt{(1+x)^2 + y^2}}.$$

The expression under the square root can be simplified: $(1 + x)^2 + y^2 = (1 + 2x + x^2) + y^2 = 1 + 2x + (x^2 + y^2) = 1 + 2x + 1 = 2 + 2x = 2(1 + x)$. Putting this back into the equation above, we get

$$\cos \theta/2 = \frac{1 + x}{\sqrt{2(1 + x)}},$$

which, after removing the square root from the denominator, becomes

$$\cos \theta/2 = \sqrt{\frac{1 + x}{2}} = \sqrt{\frac{1 + \cos\theta}{2}}.$$

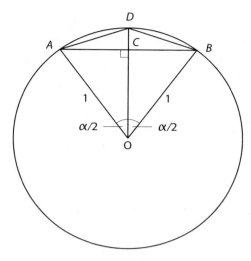

Figure 6.2. Deriving Viète's product

This is the familiar half-angle formula for the cosine (note that in proving it, we have used the Pythagorean theorem twice). From this formula and its companion, $\sin\theta/2 = \sqrt{\frac{1-\cos\theta}{2}}$ one can derive a host of related identities, including the double-angle formula for the sine, $\sin 2\theta = 2 \sin\theta \cos\theta$.[3]

We are now ready to prove Viète's product. In fig. 6.2, AB represents one side of a regular n-gon inscribed in a unit circle with center O. Let OC be the perpendicular bisector of AB, and let its extension meet the circle at D; therefore, AD and BD are each one side of a regular $2n$-gon. Let $\angle AOB = \alpha$, so $\angle AOC = \alpha/2$. Viète expressed the area $A(2n)$ of the $2n$-gon in terms of the area $A(n)$ of the n-gon. We have

$$A(n) = n \times area\ \triangle AOB = n \times twice\ area\ \triangle AOC$$

$$= 2n \times (AC \times OC)/2 = n \times AC \times OC.$$

But in the right triangle AOC we have $AC = OA \times \sin\alpha/2 = 1 \times \sin\alpha/2$ and $OC = OA \times \cos\alpha/2 = 1 \times \cos\alpha/2$, so

$$A(n) = n \sin\frac{\alpha}{2} \cos\frac{\alpha}{2}. \tag{1}$$

To find $A(2n)$, we simply replace n in equation (1) by $2n$, and α by $\alpha/2$:

$$A(2n) = 2n \sin\frac{\alpha}{4} \cos\frac{\alpha}{4}$$

$$= n \sin\frac{\alpha}{2}, \tag{2}$$

where we used the double-angle formula for the sine. Combining equations (1) and (2), we get

$$A(2n) = \frac{A(n)}{\cos \alpha/2},$$ (3)

which says that each time we double the number of sides, the area of the inscribed polygon is increased by the factor $\frac{1}{\cos \alpha/2}$.

We can reverse equation (3) to read

$$A(n) = A(2n) \cos \alpha/2$$

and apply it again and again by repeatedly doubling the number of sides:

$$A(n) = A(2n) \cos \alpha/2$$

$$= A(4n) \cos \alpha/4 \cdot \cos \alpha/2$$

$$= A(8n) \cos \alpha/8 \cdot \cos \alpha/4 \cdot \cos \alpha/2$$

$$\cdots$$

$$= A(2^k n) \cos \alpha/2 \cdot \cos \alpha/4 \cdot \ldots \cdot \cos \alpha/2^k,$$ (4)

where in the last step we wrote the chain of cosines in reverse, starting with $\cos \alpha/2$.

Viète then implemented this procedure by beginning with an inscribed square, for which $n = 4$, $A(4) = (\sqrt{2})^2 = 2$, and $\alpha = 360°/4 = 90°$. Starting with $\cos \alpha/2 = \cos 45° = \frac{\sqrt{2}}{2} = \sqrt{\frac{1}{2}}$ and repeatedly applying the half-angle formula

gives us $\cos \alpha/4 = \sqrt{\frac{1+1/\sqrt{2}}{2}} = \sqrt{\frac{1}{2} + \frac{1}{2}\sqrt{\frac{1}{2}}}$, $\cos \alpha/8 = \sqrt{\frac{1}{2} + \frac{1}{2}\sqrt{\frac{1}{2} + \frac{1}{2}\sqrt{\frac{1}{2}}}}$, and so on. Putting all this back into equation (4), we get

$$2 = A(2^k \cdot 4) \cdot \sqrt{\frac{1}{2}} \cdot \sqrt{\frac{1}{2} + \frac{1}{2}\sqrt{\frac{1}{2}}} \cdot \ldots \cdot \sqrt{\frac{1}{2} + \frac{1}{2}\sqrt{\frac{1}{2} + \ldots + \frac{1}{2}\sqrt{\frac{1}{2}}}},$$ (5)

where the last term has k nested roots.

At this point, we let the number of sides k increase beyond bound; that is, we let $k \to \infty$. But then the inscribed polygons become indistinguishable from the circle, whose area is π. Equation (5) then becomes

$$2 = \pi \sqrt{\frac{1}{2}} \cdot \sqrt{\frac{1}{2} + \frac{1}{2}\sqrt{\frac{1}{2}}} \cdots,$$

which, after dividing both sides by π, is Viète's product. Interestingly, Viète regarded $2/\pi$ as a ratio of *areas*—the area of a square to that of its circumscribing circle. Today, of course, we think of $2/\pi$ as a ratio of linear quantities (twice the diameter-to-circumference ratio of a circle).

Viète's product is regarded as the first truly analytic expression for π. Up until his time, all attempts to express π were essentially verbal, "do such and such" instructions. Archimedes, of course, already knew that in order to find the value of π, the number of sides of the inscribed and circumscribing polygons would have to be doubled indefinitely, but he carefully avoided any explicit reference to infinity, saying instead that the process could be repeated as many times as needed, until the desired accuracy was achieved. It was Viète who had the audacity to confront infinity head on with the word "etc." at the end of his formula.[4]

Even today, more than four hundred years after its discovery, Viète's product is regarded as one of the most beautiful formulas in mathematics; regrettably, it remains almost the only surviving reference to his work in modern textbooks. But when you admire its rhythmic succession of square roots of 2, remember that they are the distant ghosts of the Pythagorean theorem.

Notes and Sources

1. W. W. Rouse Ball, *A Short Account of the History of Mathematics* (1908; rpt. New York: Dover, 1960), p. 230.

2. For a more detailed account of Viète's life, see *Trigonometric Delights*, pp. 56–62.

3. To show this, multiply together the two half-angle formulas:

$$\sin\theta/2 \cdot \cos\theta/2 = \sqrt{\frac{1-\cos\theta}{2}} \cdot \sqrt{\frac{1+\cos\theta}{2}}$$

$$= \sqrt{\frac{1-\cos^2\theta}{4}} = \sqrt{\frac{\sin^2\theta}{2}} = \frac{\sin\theta}{2}.$$

Multiplying this equation by 2 and replacing $\theta/2$ with θ, we get the required formula. (Note that in the interval $0° \leq \theta \leq 360°$, $\sin\theta/2$ is always nonnegative, while $\cos\theta/2$ and $\sin\theta$ have the same sign. Consequently we can write each side of the last equation with a positive sign.)

4. We should mention that Viète's product can be obtained more easily from a little-known trigonometric identity discovered by Euler:

$$\frac{\sin x}{x} = \cos\frac{x}{2} \cdot \cos\frac{x}{4} \cdot \cos\frac{x}{8} \cdot \dots ,$$

which converges for all x. Putting $x = \pi/2$ and repeatedly using the half-angle formula for the cosine, Viète's product is obtained immediately. See my article "A Remarkable Trigonometric Identity," *Mathematics Teacher*, May 1977, pp. 452–455.

From the Infinite to the Infinitesimal

The method of Fluxions [the differential calculus]
is the general key by help whereof the modern
mathematicians unlock the secrets of
Geometry. . . . And it hath enabled them so
remarkably to outgo the ancients in discovering
theorems and solving problems.
—George Berkeley, *The Analyst* (1734)

The years 1666–1676 were an epochal decade in the history of mathematics. Working independently on either side of the English Channel, Isaac Newton (1642–1727), in England, and Gottfried Wilhelm Leibniz (1646–1716), while on a visit to Paris, were putting the finishing touches on their newly invented differential and integral calculus, the single most important event in mathematics since Euclid wrote his *Elements* nearly two thousand years earlier.

Newton, ever the physicist, cast his calculus in dynamic terms. He thought of a variable as a kind of fluid in a continuous state of motion—he called it a *fluent*—and a function as a relation between two fluents, each with its own rate of flow. He defined the derivative of a function—its *fluxion*, as he called it—as the ratio between the two rates of flow, each with respect to time. This ratio was a measure of the rate of change of one variable with respect to the other—our modern derivative.

Leibniz, a man of many interests but first and foremost a philosopher, followed a more abstract approach. He thought of a function as an equation relating one variable, say x, to another variable, y; today we write such a relation as $y = f(x)$. Leibniz then allowed x and y to increase by small amounts dx and dy, respectively. The ratio of these increments, $\frac{dy}{dx}$, is a measure of the rate of change of y with respect to x. This is illustrated in figure 7.1, which shows the graph of a function $y = f(x)$ and a point P on it. Draw the tangent line to the graph at P and consider a neighboring point T, also on the tangent line. This produces the small triangle PRT, which Leibniz called the *characteristic triangle*. Its sides PR and RT are the increments in the x and y coordinates as we move from P to T, that is, dx and dy. Leibniz argued that if these increments

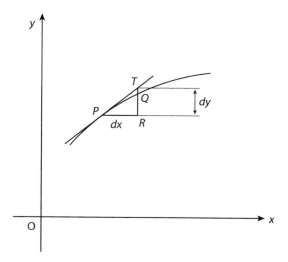

Figure 7.1. The characteristic triangle

are sufficiently small, the straight line segment PT will be nearly identical with the *curved* segment PQ; that is, the tangent line between P and T will nearly coincide with the graph *at P*. Consequently, the slope of the tangent line is a measure of the rate of change of the function at P. This slope is simply the ratio $\frac{dy}{dx}$; today we call this ratio—or more precisely, its limit as dx and dy tend to zero—the *derivative* of $f(x)$ and denote it by $f'(x)$.

We note that as P moves along the graph, the characteristic triangle changes continually, and with it the ratio $\frac{dy}{dx}$; that is, the derivative changes its value from one point to another and is therefore itself a function of x (hence the notation $f'(x)$ and the name *derivative*, short for *derivative function*). The process of finding the derivative is called *differentiation*.

Leibniz sometimes thought of dx and dy as small but finite quantities; at other times he thought of them as infinitely small, or *infinitesimals*; today we call them *differentials*. Regardless, Leibniz's notation enabled him to deal with the quantity $\frac{dy}{dx}$ *as if* it were a ratio of two ordinary quantities (even if they were infinitely small). In this sense, most of the operating rules of the differential calculus—the staple of a first-semester calculus course—are little more than algebraic manipulations of differentials.

Leibniz followed a similar approach to tackle another problem, finding the area under the graph of a function. He again thought of dx as a small increment in x. He divided the area under the graph into many narrow vertical strips, each of height $y = f(x)$ and width dx (fig. 7.2). The area of a typical strip is then $y\,dx = f(x)dx$. By taking the sum of all these areas, Leibniz got an expression for the total area A under the graph between the points $x = a$ and

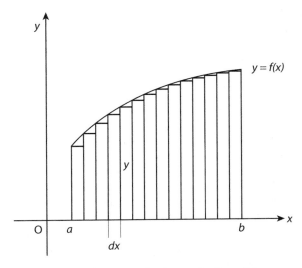

Figure 7.2. Approximating the area under the graph of $y = f(x)$

$x = b$. This process is called *integration*; we write $A = \int_a^b f(x)dx$, read *the definite integral of $f(x)$ from a to b*. The procedure can be modified to find other quantities associated with the function, such as the surface area and volume of the solid obtained by revolving the graph of $f(x)$ about the x-axis.

These two problems—finding the slope, or rate of change, of a function $f(x)$ at a given point on it, and finding the area under the graph of $f(x)$ over a given interval—are the backbone of the differential and integral calculus. At first glance, the two seem to be unrelated, but Newton and Leibniz showed that they are, in fact, *inverse problems*. Specifically, to find the area under the graph of $f(x)$, we must first find an *antiderivative* of $f(x)$—a function $F(x)$ whose derivative is $f(x)$. The value of $\int_a^b f(x)dx$ is then the difference between the values of $F(x)$ at $x = b$ and $x = a$. That is,

$$\int_a^b f(x)dx = F(b) - F(a), \quad \text{where } F'(x) = f(x)$$

(the expression $F(b) - F(a)$ is often written as $F(x) \big|_a^b$). This inverse relation is known as the *fundamental theorem of calculus*.

Newton and Leibniz, following their separate approaches, went on to develop the operating rules of their new calculus, transforming it into a powerful tool that would affect practically every branch of mathematics and science. But their feat had an ugly aftermath: the erstwhile colleagues became bitter enemies, each accusing the other of stealing his invention. The dispute would last long after the two protagonists were dead; it would poison the scientific atmosphere in Europe for over a century and stifle nearly all further development of the subject in England. Today, at last, Newton and Leibniz are given equal credit as the co-inventors of the calculus.[1]

Once the operating rules of the new calculus were set, the door was open to tackle a vast range of problems whose solution had been out of reach for centuries. One of the first to be attempted was finding the length of a curve between two points on it. This process is known as *rectification*. Since antiquity it had been believed that a curved line cannot be rectified—that its length could never be made equal to the length of a straight line segment. This view was shared even by Fermat and Pascal; the latter expressed his amazement that nature allows us to find the *area* under certain curves, but not their length. The invention of calculus would prove him wrong.

Of course, to find the length of a straight-line segment between the points (x_1, y_1) and (x_2, y_2), we simply use the distance formula $s = \sqrt{(x_1 - x_2)^2 + (y_1 - y_2)^2}$, where we have replaced the usual letter d for distance by s, to avoid confusing it with the d in $\frac{dy}{dx}$. But for all other curves, we must follow a "local" approach. We divide the graph of $f(x)$ into many small straight-line segments (fig. 7.3). Each segment is the hypotenuse of a characteristic triangle with sides dx and dy. If this triangle is small enough, we can regard its hypotenuse as a close approximation to an element of *arc length, ds*, of the graph; that is, for sufficiently small dx and dy, we have the approximate equation

$$ds^2 = dx^2 + dy^2, \tag{1}$$

which is the differential version of the Pythagorean theorem.

Solving this equation for ds, we get $ds = \sqrt{dx^2 + dy^2} = \sqrt{1 + (dy/dx)^2}\,dx$. The total arc length s along the curve is the "infinite sum" of all the ds; that is, the definite integral

$$s = \int_a^b \sqrt{1 + (dy/dx)^2}\,dx = \int_a^b \sqrt{1 + y'^2}\,dx, \tag{2}$$

where $y = f(x)$ is the equation of the curve, $y' = \frac{dy}{dx}$ is its derivative, and a and b are the endpoints of the interval of integration.

With the help of equation (2) we can find, in principle, the arc length of any curve whose equation is given in the form $y = f(x)$, provided the integral in equation (2) exists (has a finite value). In practice, however, even simple

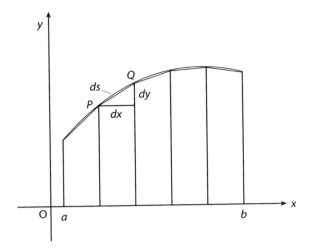

Figure 7.3. Rectification of a curve

functions may make the actual integration a formidable task. As an example, let us find the arc length of the parabola $y = x^2$ from $x = 0$ to $x = a$. We have $y' = 2x$, so $s = \int_0^a \sqrt{1 + (2x)^2}\, dx$. This simple-looking integral may tax the patience of many a first-semester calculus student; the result is $\frac{1}{2} a\sqrt{1 + 4a^2} + \frac{1}{4} \ln(2a + \sqrt{1 + 4a^2})$, where $\ln$ stands for natural logarithm.

The rectification of curves presented a considerable challenge to the early pioneers of the calculus, when new techniques of integration were constantly being invented. Among the first to be rectified were two curves whose equations are usually given in nonrectangular coordinates: the *logarithmic spiral* and the *cycloid*. The former is famous for its frequent occurrence in art and nature; a nautilus shell has the exact shape of a logarithmic spiral, as does the arrangement of seeds in a sunflower. The equation of the spiral is expressed in *polar coordinates*, in which a point P is located in terms of its distance r from the origin O (the "pole") and the angle θ between the positive x-axis and the radius vector OP, measured counterclockwise in radians. The polar equation of the spiral is $r = e^{a\theta}$, where e is the base of natural logarithm (about 2.71828) and a is a constant that determines the rate of growth of the spiral; if $a > 0$, r increases as we turn the angle θ counterclockwise, resulting in a left-handed spiral (fig. 7.4a); if $a < 0$, r decreases and we get a right-handed spiral (fig. 7.4b). As the spiral winds inward toward the pole, it turns around it indefi-

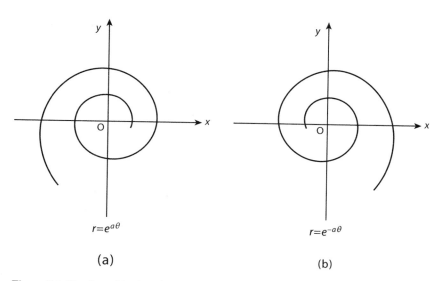

$r=e^{a\theta}$

(a)

$r=e^{-a\theta}$

(b)

Figure 7.4. Two logarithmic spirals: (a) left-handed; (b) right-handed

nitely, getting ever closer to but never reaching it. Imagine therefore the surprise when the Italian Evangelista Torricelli in 1645 showed that the total distance from any point on the spiral to the pole is finite!

Torricelli (1608–1647), a disciple of Galileo and briefly his assistant, is best known for his invention of the mercury barometer. But like most scientists of his day, his work covered a wide range of subjects, including mathematics. He was particularly interested in the properties of various curves, among them the cycloid (to be discussed shortly) and the logarithmic spiral. The integral calculus was still two decades in the future, so when Torricelli attempted to rectify the spiral, he had to use a roundabout method. He took advantage of a unique property of the spiral: any part of it, no matter how large or small, looks exactly like any other part (today we call such a self-similar curve a *fractal*). This means that if we turn the radius vector OP in equal amounts, its length r increases in equal *ratios*—it follows a geometric progression. So Torricelli divided the spiral into many narrow sectors of equal angular width $d\theta$. Each sector forms a characteristic triangle with sides dr and $rd\theta$, as shown in figure 7.5. If this triangle is sufficiently small, we can replace the element of arc ds (arc PQ in the figure) with the line segment PQ. By the Pythagorean theorem we then have

$$ds = \sqrt{(dr)^2 + (rd\theta)^2} = \sqrt{r^2 + (dr/d\theta)^2}\, d\theta. \qquad (3)$$

Summing up these arc-length elements and using the property just mentioned, Torricelli found that the total arc length from P to the pole O is equal to the length of the tangent line to the spiral at P, taken between P and the y-axis

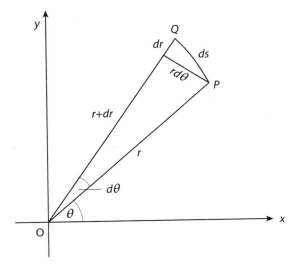

Figure 7.5. An element of arc length in polar coordinates

(fig. 7.6). This was the first known rectification of a transcendental (nonalgebraic) curve; a proof is given in Appendix F.[2]

❖   ❖   ❖

The cycloid is the curve formed by a point on the rim of a wheel rolling without sliding along a straight line. It is easiest to express it by a pair of parametric equations. Let a be the radius of the generating circle, $P(x, y)$ a point on the circumference, and θ the angle through which the circle has turned, measured clockwise in radians from the point where P touches the x-axis (fig. 7.7). As the circle rolls along the x-axis, P describes an arc of length $a\theta$ along the circumference, and this arc is transformed into a linear motion along the x-axis. The x-coordinate of P is therefore $a\theta - a \sin \theta = a(\theta - \sin \theta)$, and the y-coordinate is $a - a \cos \theta = a(1 - \cos \theta)$:

$$x = a(\theta - \sin \theta), \quad y = a(1 - \cos \theta). \tag{4}$$

Like the logarithmic spiral, the cycloid has many interesting properties; for example, when turned upside down, it is the curve along which a particle under the force of gravity will slide down in the least possible time (and also the curve along which a particle will oscillate with a constant period, regardless of the amplitude of oscillations). So it is no wonder that many mathematicians in the seventeenth and eighteenth centuries studied this curve, among them Galileo, Descartes, Torricelli, Huygens, and the Bernoulli brothers. Torricelli is credited with the first published demonstration that the *area* under one arch

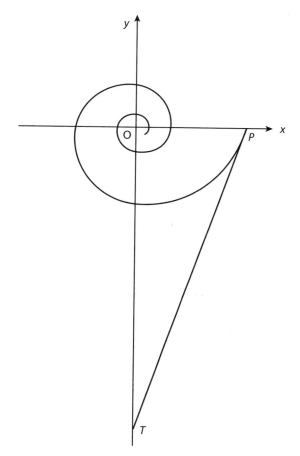

Figure 7.6. Rectification of the logarithmic spiral

of the cycloid is $3\pi a^2$, that is, three times the area of the generating circle (the Frenchman Gilles Persone Roberval preceded him by a few years but failed to publish his results, triggering a bitter priority dispute between the two). But it was the renowned English architect Sir Christopher Wren (1632–1723) who first found its arc length.

Wren's name is remembered today mainly for rebuilding London after the Great Fire of 1666, a monumental task whose crowning achievement was the magnificent St. Paul's Cathedral. But Wren was also a mathematician, holding the chair of Savilian professor of astronomy at Oxford. Among his discoveries was the fact—still causing amazement when students first learn of it—that the saddle-shaped surface of a hyperboloid (the surface generated when the hyperbola $x^2 - y^2 = 1$ revolves about the y-axis) can be constructed entirely from two families of straight lines, where each line of one family crosses every line

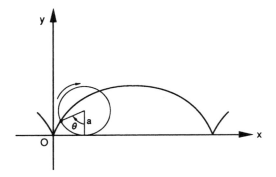

Figure 7.7. Cycloid

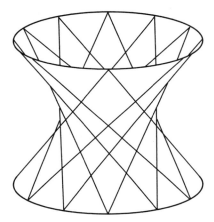

Figure 7.8. Hyperboloid of revolution

of the other, but lines from the same family do not cross (fig. 7.8). His rectification of the cycloid dates to 1658: he found that the length of each arch is exactly 8a (for a proof, see Appendix F); surprisingly, the result is independent of π.[3]

❖   ❖   ❖

We consider two more curves whose rectification is relatively easy: the *astroid* and the *catenary*. The astroid is the curve generated when a rod of unit length whose endpoints slide along the x- and y-axes assumes all possible positions (fig. 7.9). Its equation, in implicit form, is $x^{2/3} + y^{2/3} = 1$. Solving this equation for y gives us the explicit equation, $y = (1 - x^{2/3})^{3/2}$, from which we can find y' and put it in the formula $s = \int_0^1 \sqrt{1 + y'^2}\, dx$. Working out this inte-

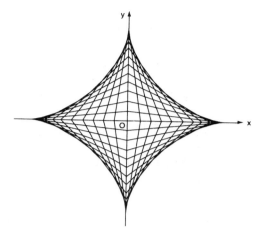

Figure 7.9. Astroid

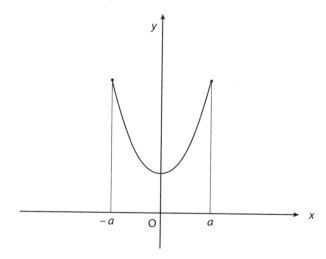

Figure 7.10. Catenary

gral, we find that the arc length of one-quarter of the astroid is 3/2, that is, $1\frac{1}{2}$ times the length of the generating rod (for the details, see Appendix F). We will encounter the astroid again in chapter 10.[4]

The catenary is the curve formed by a chain of uniform thickness hanging freely under the force of gravity between two fixed points (the word comes from the Latin *catena*, a chain). The shape of the catenary (fig. 7.10) stirred up an intense debate among mathematicians of the seventeenth century. Galileo, for example, believed that the catenary is a parabola, which it certainly resembles.

But in 1691, in response to a challenge by Jakob Bernoulli, three mathematicians independently came up with the correct answer; they were Jakob himself, his younger brother Johann, and Leibniz. The curve turned out to have the equation $y = \dfrac{e^x + e^{-x}}{2}$, where e is again the base of natural logarithms. The expression $\dfrac{e^x + e^{-x}}{2}$ is called the *hyperbolic cosine of x* and is written $\cosh x$; similarly, we define the *hyperbolic sine of x* as $\dfrac{e^x - e^{-x}}{2}$, written $\sinh x$. The equation of the catenary can now be written as $y = \cosh x$.

To find the arc length of the catenary, we take advantage of certain formal similarities between the hyperbolic functions and the circular (trigonometric) functions $\sin x$ and $\cos x$. For example, corresponding to the identity $\cos^2 x + \sin^2 x = 1$—the trigonometric version of the Pythagorean theorem—we have the hyperbolic identity $\cosh^2 x - \sinh^2 x = 1$ (note, however, the negative sign in the second term).[5] Equally remarkable is the similarity between the derivatives of these functions; we have $(\sin x)' = \cos x$ and $(\cos x)' = -\sin x$, while the analogous hyperbolic formulas are $(\sinh x)' = \cosh x$ and $(\cosh x)' = \sinh x$ (both with positive signs).

These features make the rectification of the catenary particularly simple. Starting with $y = \cosh x$, we have $\sqrt{1 + y'^2} = \sqrt{1 + \sinh^2 x} = \sqrt{\cosh^2 x} = \cosh x$. Putting this into equation (2), we find that the length of the catenary, taken between the points $x = -a$ and $x = a$ (symmetrically positioned about the lowest point of the curve) is $s = \displaystyle\int_{-a}^{a} \cosh x\, dx = \sinh x \Big|_{-a}^{a} = \sinh a - \sinh(-a) = 2 \sinh a$, where we have used the fact that $\sinh x$ is an *odd* function, that is, $\sinh(-x) = -\sinh x$. Interestingly, $\displaystyle\int_{-a}^{a} \cosh x\, dx$ also represents the *area* under the catenary from $x = -a$ to $x = a$, so the arc length of the catenary is numerically equal to the area under it.[6]

One might be tempted to conclude that it would be equally easy to rectify the trigonometric cosine function, but this is not the case. Following the same steps as we took above but this time with $y = \cos x$, we end up with the expression $\sqrt{1 + \sin^2 x}$, whose antiderivative cannot be expressed in terms of the elementary functions.[7] This is due to the "plus" sign in the trigonometric identity $\cos^2 x + \sin^2 x = 1$, which makes it impossible to simplify the expression $1 + \sin^2 x$. We can, of course, find the approximate value of $\displaystyle\int \sqrt{1 + \sin^2 x}\, dx$, say from $x = 0$ to $x = \pi$ (half the cycle of $\cos x$) by numerical integration; the result, as computed by a TI-83 graphing calculator, is about 3.82. But we cannot write the outcome as an analytic expression in closed form.

❖ ❖ ❖

We have come a long way since Euclid formulated the Pythagorean theorem as Proposition I 47 of the *Elements*. First conceived as a purely geometric relation between the areas of the squares built on the sides of a right triangle, it gradually evolved into an algebraic relation between the lengths of the three sides, to be used whenever two sides are given and the third is to be found. Hardly could Pythagoras have imagined that his theorem would one day be used to find the length of almost *any* curve, if we only know its equation.[8] For this we have to thank the very idea that the Greeks had barred so adamantly from their world, the idea of infinity.[9]

Notes and Sources

1. Needless to say, this scant sketch of the calculus is not nearly enough even as a crude introduction to the subject. For a more detailed outline, see *e: the Story of a Number*, chaps. 8 and 9.

2. For more on the logarithmic spiral and its role in art and nature, see Theodore Andrea Cook, *The Curves of Life: Being an Account of Spiral Formations and Their Application to Growth in Nature, to Science and to Art* (1914; rpt. New York: Dover, 1979); Matila Ghyka, *The Geometry of Art and Life* (1946; rpt. New York: Dover, 1977); Jay Hambidge, *The Elements of Dynamic Symmetry* (1926; rpt. New York: Dover, 1967); D'Arcy W. Thompson, *On Growth and Form* (1917; rpt. London and New York: Cambridge University Press, 1961); and *To Infinity and Beyond*, chaps. 11 and 21.

3. For more on the cycloid, see Robert C. Yates, *Curves and Their Properties* (Reston, Va.: National Council of Teachers of Mathematics, 1974), pp. 65–70.

4. For more on the astroid, see ibid., pp. 1–3.

5. It is from this identity that the term "hyperbolic functions" is derived. For, if we write $x = \cosh t$, $y = \sinh t$, where t is a parameter, then it is easy to prove that $x^2 - y^2 = 1$, showing that the point (x, y) lies on the hyperbola $x^2 - y^2 = 1$. This is analogous to the pair of equations $x = \cos t$, $y = \sin t$, whose corresponding point (x, y) lies on the unit circle $x^2 + y^2 = 1$ (thus the term "circular functions").

6. For more on the catenary, see Yates, *Curves*, pp. 12–14, and *e: The Story of a Number*, chap. 12.

7. The elementary functions consist of power functions, polynomials and ratios of polynomials, exponential, trigonometric and hyperbolic functions and their inverses, and any finite combination of these functions obtained by addition, subtraction, multiplication, division, exponentiation, and root extraction.

8. I say "almost," because there exist "pathological" curves for which the arc length does not have a finite value (even over an interval where the function is completely defined). As an example, consider the function $f(x) = x \sin \frac{1}{x}$, whose graph oscillates between the lines $y = \pm x$. This function is not defined at $x = 0$, but we can assign it the value $f(0) = 0$, resulting in a continuous function for all values of x. As $x \to 0$, the frequency of oscillations increases without bound. As a result, the arc length of the graph from $x = 0$ to any other point is infinite.

9. For more on the early history of rectification, see Margaret E. Baron, *The Origins of the Infinitesimal Calculus* (1969; rpt. New York: Dover, 1987), pp. 223–228.

A Remarkable Formula by Euler

Euler calculated without any apparent effort,
just as men breathe, as eagles sustain themselves
in the air.
  —François Arago (1786–1853)

In 1734 the Swiss mathematician Leonhard Euler (1707–1783) solved one of the outstanding problems of his time: to find the sum of the infinite series $1 + 1/2^2 + 1/3^2 + \ldots$. Jakob Bernoulli had already proved in 1689 that this series converges, but no one could find its exact sum. Among the many mathematicians who tried but failed were the Bernoulli brothers, Johann and Jakob. Euler, using methods that today would not be acceptable even from a beginning calculus student, dropped a bombshell when he announced that the series converges to $\pi^2/6$ (approximately 1.64493). His method was flawed, but he got away with it: he found the correct sum.[1]

Euler's discovery is remarkable because of the unexpected appearance of π in a series that involves only the natural numbers.[2] Moreover, the fact that it consists of the sum of *squares* makes one wonder if it has anything to do with the Pythagorean theorem. And indeed it does. In figure S3.1, let $r_1 = OP_1$ be the line segment from 0 to 1 along the x-axis. At P_1 erect P_1P_2 perpendicular to the x-axis and of length 1/2. The radius vector from O to P_2 has length $r_2 = \sqrt{1 + 1/2^2}$. At P_2 erect P_2P_3 perpendicular to OP_2 and of length 1/3. The length of OP_3 is $r_3 = \sqrt{1 + 1/2^2 + 1/3^2}$. Continuing in this manner n times, we arrive at the point P_n whose distance from O is $r_n = \sqrt{1 + 1/2^2 + 1/3^2 + \ldots + 1/n^2}$. As n increases, the sum under the square root slowly increases and approaches $\pi^2/6$ as $n \to \infty$. Consequently, P_n slowly spirals outward, getting ever closer to a limiting circle of radius $r_\infty = \pi/\sqrt{6} \sim 1.28255$. At the same time, the total length of the segments $OP_1, P_1P_2, \ldots, P_{n-1}P_n$ tends to infinity, as follows from the fact that the harmonic series $1 + 1/2 + 1/3 + 1/4 + \ldots$ diverges.

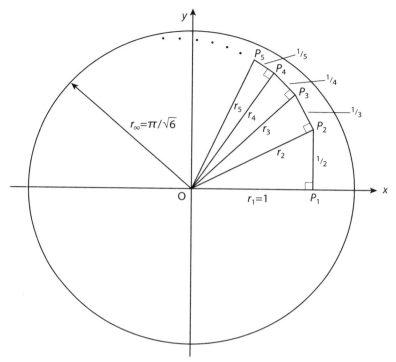

Figure S3.1. Geometric representation of the series of $\displaystyle\sum_{n=1}^{\infty}\frac{1}{n^2}$

Note that as long as n is finite, we can implement this process with a straightedge and compass, as each of the steps involved is constructible with these tools.[3] So, in principle we may use the process to approximate the value of π graphically. However, before you try to do so, be warned that it will be a rather long and tedious process, as the rate of convergence of Euler's series is very slow: it takes 628 terms to find π correct to just two places, that is 3.14!

Euler did not stop with the series $1 + 1/2^2 + 1/3^2 + \dots$. Using similar methods, he was able to find the sum of the series $1 + 1/2^k + 1/3^k + \dots$ for all even values of k from 2 to 26; for this last value he found the sum to be

$$\frac{2^{24} \times 76{,}977{,}927 \times \pi^{26}}{1 \times 2 \times 3 \times \dots \times 27}.$$

The same series for *odd* values of k, however, is much more difficult to handle, and until very recently, the nature of the sum for $k = 3$ was not known.[4]

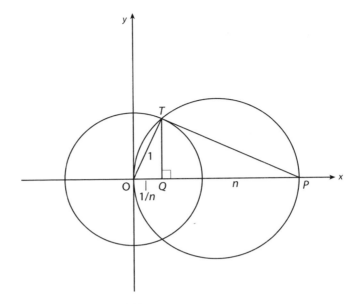

Figure S3.2. Construction of 1/*n*

Notes and Sources

1. Basically, Euler applied the rules of ordinary, finite algebra to infinite series (specifically, the power series for sin *x*; see p. 120). This is not always allowed and might even lead to absurd results, but in this case Euler got the correct answer. As to how he found it, see William Dunham, *Journey through Genius: The Great Theorems of Mathematics* (New York: John Wiley, 1990), chap. 9. See also *Trigonometric Delights*, pp. 156–161.

2. It was not, however, the first such series to be discovered. In 1671, James Gregory (1638–1675) used the recently invented integral calculus to show that $1 - 1/3 + 1/5 - 1/7 + - \ldots$ converges to $\pi/4$. The same series was independently discovered by Leibniz in 1674 and is known as the *Gregory-Leibniz series*.

3. To construct 1/n, allocate a line segment *OP* of length *n* along the *x*-axis (fig. S3.2). Draw a circle with *OP* as diameter, and let this circle intersect the unit circle at *T*. Drop the perpendicular from *T* to the *x*-axis, meeting it at *Q*. By a well-known theorem (Euclid III, 20), $\angle OTP$ is a right angle. Therefore, triangles *OQT* and *OTP* are similar, and so $OQ/OT = OT/OP$. But $OT = 1$, and so $OQ = 1/OP = 1/n$.

4. It is known, however, that the series $1 + 1/2^k + 1/3^k + \ldots$ converges for all $k > 1$ and diverges for $k \le 1$ (the case $k = 1$ is the harmonic series). For the case $k = 3$, see Alfred Van der Poorten, "A Proof That Euler Missed . . . —Apéry's Proof of the Irrationality of $\zeta(3)$," *Mathematical Intelligencer*, 1 (1979), pp. 195–203. Apéry in 1978 proved that the sum of this series is approximately 1.202.

The series $1 + 1/2^k + 1/3^k + \ldots$, when regarded as a function of the exponent k (where k can assume complex values) is known as the *zeta function* and denoted by $\zeta(k)$. It is the subject of intense interest in mathematics because of the *Riemann hypothesis*, which says that all complex zeros of the zeta function lie on the vertical line $x = 1/2$ of the complex plane. Despite numerous attempts, the conjecture remains unproved. See John Derbyshire, *Prime Obsessions: Bernhard Riemann and the Greatest Unsolved Problem in Mathematics* (Washington, D.C.: Joseph Henry Press, 2003), and Dan Rockmore, *Stalking the Riemann Hypothesis: The Quest to Find the Hidden Law of Prime Numbers* (New York: Pantheon Books, 2005).

371 Proofs, and Then Some

The Pythagorean Theorem is regarded as the most
fascinating Theorem of all of Euclid, so much so,
that thinkers from all classes and nationalities,
from the aged philosopher in his armchair to the
young soldier in the trenches next to no-man's-
land, 1917, have whiled away hours seeking a new
proof of its truth.

—Elisha Scott Loomis, *The Pythagorean Proposition*

Elisha Scott Loomis (1852–1940) is not a household name in mathematical circles; as far as I know, there is no equation or theorem named after him, and the few books he wrote are largely forgotten today. Except one: his 285-page *The Pythagorean Proposition*,[1] in which he collected and classified 371 proofs of the Pythagorean theorem.

Loomis, as described in the preface to one of his publications, was a "philosopher, mathematician, author, genealogist, and civil engineer, [but] prized more than any thing else the title of 'Teacher,' for it was in his fifty years as a teacher that he exerted the deepest influence on all who came in contact with him."[2] He was born in Medina County, Ohio, the eldest of eight children and an eighth-generation descendant of Joseph Loomis, a native of England who emigrated to America in 1639; among his many family members were several army officers and one mathematician, Elias Loomis (1811–1899). Having been left fatherless at the age of twelve, Elisha went to work to support his family. But he was an assiduous student, showing an early inclination toward mathematics; on one occasion he walked seven miles to a nearby town to purchase an algebra textbook, which he then mastered on his own as no teacher at his school was competent enough to guide him. In 1880 he received his B.Sc. from Baldwin University, to be followed by an M.A. in 1886 and a Ph.D. two years later. For a while he served as chairman of the mathematics department at Baldwin, but his real passion was teaching. Leaving his academic post, he became head of the mathematics department at West High School in Cleveland, holding the post for twenty-eight years. All the while he continued his studies, earning a law degree from the Cleveland Law School in 1900, which led to his admission to the Ohio bar. And if all this were not enough, he also studied civil engineering and served as the municipal engineer for the village of Berea.

Loomis was a prolific author. He wrote over a hundred articles and published several books on subjects ranging from the teaching of geometry to ethics, philosophy and religion. He was also a meticulous recorder of his family genealogy, tracing his ancestors' whereabouts all the way back to Joseph Loomis. This resulted in an 859-page revision of *Descendants of Joseph Loomis in America and His Antecedents in the Old World* (1908), a work first published in 1875 by his cousin, the mathematician Elias Loomis; Elisha's revision listed and classified no fewer than 32,000 names. To cap it all, in 1934 he wrote his own obituary, complete with instructions on how it should be read at his funeral. His request was honored to the letter—"except for the date of his death and the addresses of some of his survivors." As befitting the occasion, he spoke of himself in the third person, saying, "In his 50 years as a teacher he plowed habit-formation grooves in the plastic brains of over 4,000 boys and girls and young men and women."[3]

But Loomis considered *The Pythagorean Proposition,* which he wrote in 1907 but did not publish until 1927, his best work. He revised it in 1940, the year of his death. Pedantic to the end, Loomis included an addendum listing several proofs that came to his attention "since June 23, 1939, the day on which I finished page 257 of this 2nd edition," adding that "they came to me from everywhere." He signed the work with "E. S. Loomis, Ph.D., at age nearly 88. May 1, 1940." In 1968 the work was reprinted by the National Council of Teachers of Mathematics, the first in a series of classics of "timeless quality" (to quote the publishers) in mathematics education.

The Pythagorean Proposition evidently reflects the idiosyncratic character of its author. Scattered throughout its pages are twelve portraits of luminaries such as Euclid, Copernicus, Descartes, Galileo, and Newton, and not surprisingly, Pythagoras himself. But the portrait adorning the frontispiece is none other than Loomis's—a commanding figure with the stern face of an authoritarian school principal (fig. 8.1). The opening page of the text shows a cryptic triangle with its vertices marked with the letters E, S, L—evidently the author's initials—and an enigmatic figure 4 with the inscription "32°" above it (fig. 8.2). This is followed by a lengthy biography of Pythagoras, which, like all such biographies, must be viewed with a great deal of skepticism. Loomis comments that in the Middle Ages, it was required of a student taking his master's degree in mathematics to offer a new and original proof of the Pythagorean theorem; this, Loomis believed, accounts for the large number of proofs that have been proposed over the years. As for Pythagoras's own proof, Loomis said: "Whether his proof of the famous theorem was wholly original no one knows; but we now know that geometers of Hindustan knew this theorem centuries before his time; whether he knew what they knew is also un-

Figure 8.1. Elisha Scott Loomis

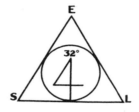

GOD GEOMETRIZES
CONTINUALLY~*PLATO.*

Figure 8.2. First page of *The Pythagorean Proposition*

known. But he, of all the masters of antiquity, carries the honor of its place and importance in our Euclidean Geometry."

Loomis divided the 371 proofs into two broad classes: algebraic and geometric, to which he added two smaller ones, "quaternionic" ("quaternions" being his word for vectors), and "dynamic" proofs based on mechanical principles. His distinction between "algebraic" and "geometric" proofs is not always clear, but it seems to be based on whether the proof shows that $c^2 = a^2 + b^2$ (regarded as a purely algebraic statement), or whether it compares the areas of the squares built on the hypotenuse and the two sides, the way Pythagoras had perceived it. The 109 algebraic proofs are further divided into

PYTHAGOREAN MAGIC SQUARES

One

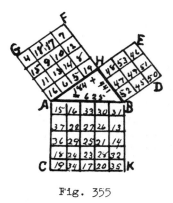

Fig. 355

The sum of any row, column or diagonal of the square AK is 125; hence the sum of all the numbers in the square is 625. The sum of any row, column or diagonal of square GH is 46, and of HD is 147; hence the sum of all the numbers in the square GH is 184, and in the square HD is 441. Therefore the magic square AK (625) = the magic square HD (441) + the magic square HG (184).

Formulated by the author, July, 1900.

Two

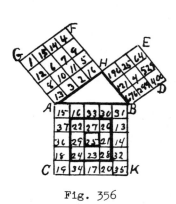

Fig. 356

The square AK is composed of 3 magic squares, 5^2, 15^2 and 25^2. The square HD is a magic square each number of which is a square. The square HG is a magic square formed from the first 16 numbers. Furthermore, observe that the sum of the nine square numbers in the square HD equals 48^2 or 2304, a square number.

Formulated by the author, July, 1900.

Figure 8.3. Two magic squares from *The Pythagorean Proposition*

seven subgroups, and the 256 geometric proofs into ten subgroups according to a variety of criteria. The book is supplemented by one "Pythagorean Curiosity" (see p. 140) and five Pythagorean magic squares, two of which we show here (fig. 8.3).

To follow even a fraction of all these proofs can test one's patience to the limit, all the more so in light of Loomis's terse style. It must also be said that many of the proofs are little more than subtle variations of one another. Still, if one digs patiently through this trove, one is bound to find some treasures. All the classical proofs are there, as well as some lesser-known ones by luminaries like Huygens and Leibniz. There is a proof by a Miss E. A. Coolidge, a blind girl, dating back to 1888, and another by Miss Ann Condit, a sixteen-year-old high school student. There is one proposed by a future American president, and yet another attributed to Leonardo da Vinci; in short, a gallery of the famous and not-so-famous in the history of mathematics.

Following are some highlights from Loomis's book; on occasion I have edited his proofs to make them easier to follow.

The shortest proof. Loomis gives in modern notation Euclid's second proof (VI 31; see p. 41), but credits it to the French mathematician Adrien-Marie Legendre (1752–1833). Moreover, notwithstanding his claim that "this is the shortest proof possible," there is a still shorter one, as I will show in Sidebar 4.

The longest proof. Being too long to repeat here in full, I refer the reader to figure 8.4. This proof, too, is credited to Legendre.

Ptolemy's proof. Claudius Ptolemaeus, commonly known as Ptolemy (ca. 85–ca. 165 CE) was, after Archimedes, arguably the greatest applied mathematician of antiquity, and certainly the most influential. He lived in Alexandria, but like Euclid, little is known about his life (he is unrelated to the Ptolemy dynasty that ruled Egypt after the death of Alexander the Great). Ptolemy's work was chiefly in geography and astronomy; his most famous work, the *Almagest* (see p. 58), is a treatise on trigonometry and mathematical astronomy. There we find the following result, known as *Ptolemy's theorem*:

> *The rectangle contained by the diagonals of any quadrilateral inscribed in a circle is equal to the sum of the rectangles contained by the pairs of opposite sides.*

To understand this cryptic statement, we must again remember that the Greeks regarded a product $a \times b$ of two numbers as the area of a rectangle with sides of length a and b. Thus "the rectangle contained by the diagonals of any quadrilateral inscribed in a circle" means the area of a rectangle whose sides are the diagonals of an inscribed quadrilateral, with a similar interpretation for "the rectangle contained by the pair of opposite sides"; in short, "a rectangle contained by" simply means "a product of." Ptolemy's theorem can then be formulated as follows: *In a quadrilateral inscribed in a circle, the product of the diagonals is equal to the sum of the products of the opposite sides.* Referring to figure 8.5, this means that

$$AC \times BD = AB \times CD + BC \times DA. \tag{1}$$

Ninety

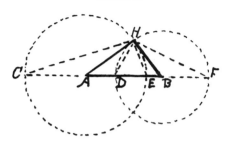

$AH^2 = AD(AB + BH).---(1)$ $BH^2 = BE(BA + AH).---(2)$
$(1) + (2) = (3)$ $BH^2 + AH^2 = BH(BA + AH) + AD(AB + BH)$
$= BH \times BA + BE \times AH + AD \times HB + AD \times BH$
$= HB(BE + AD) + AD \times BH + BE \times AH + BE \times AB - BE \times AB$

$= AB(BE + AD) + AD \times BH + BE(AH + AB) - BE \times AB$
$= AB(BE + AD) + AD \times BH + BE(AH + AE + BE) - BE \times AB$
$= AB(BE + AD) + AD \times BH + BE(BE + 2AH) - BE \times AB$
$= AB(BE + AD) + AD \times BH + BE^2 + 2BE \times AH - BE \times AB$
$= AB(BE + AD) + AD \times BH + BE^2 + 2BE \times AE - BE(AD + BD)$
$= AB(BE + AD) + AD \times BH + BE^2 + 2BE \times AE - BE \times AD$
  $- BE \times BD$
$= AB(BE + AD) + AD \times BH + BE(BE + 2AE) - BE(AD + BD)$
$= AB(BE + AD) + AD \times BH + BE(AB + AH) - BE(AD + BD)$
$= AB(BE + AD) + AD \times BH + (BE \times BC = BH^2 = BD^2)$
  $- BE(AD + BD)$
$= AB(BE + AD) + (AD + BD)(BD - BE)$
$= AB(BE + AD) + AB \times DE = AB(BE + AD + DE)$
$= AB \times AB = AB^2.$ $\therefore h^2 = a^2 + b^2.$ Q.E.D.
    a. See Math. Mo. (1859), Vol. II, No. 2, Dem.
28, fig. 13--derived from Prop. XXX, Book IV, p. 119,
Davies Legendre, 1858; also Am. Math. Mo., Vol. IV,
p. 12, proof XXV.

Figure 8.4. The longest proof in *The Pythagorean Proposition*

Proof:

Using one side, say *AB*, as the initial side, construct $\angle ABE = \angle DBC$.
Angles *CAB* and *CDB* are also equal, having the common chord *BC*.
Therefore, triangles *ABE* and *DBC* are similar, having two pairs of
equal angles. Hence *AE* / *AB* = *DC* / *DB*, from which we get

$$AE \times DB = AB \times DC. \tag{2}$$

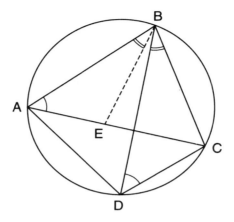

Figure 8.5. Ptolemy's theorem

If we now add $\angle EBD$ to both sides of the equation $\angle ABE = \angle DBC$, we get $\angle ABD = \angle EBC$. But angles BDA and BCA are also equal, having the common chord AB. Therefore, triangles ABD and EBC are similar, hence $AD \mathbin{/} DB = EC \mathbin{/} CB$, and so

$$EC \times DB = AD \times CB. \tag{3}$$

Finally, adding equations (2) and (3), we get $(AE + EC) \times DB = AB \times DC + AD \times CB$; and replacing $AE + EC$ by AC, we get the required result (note that all the sides are *nondirected* line segments, so $BD = DB$, etc.).[4]

The Pythagorean theorem now follows as a special case if we let quadrilateral $ABCD$ be a rectangle (fig. 8.6); then all four vertices are right angles, and furthermore $AB = CD$, $BC = DA$, and $AC = BD$. Ptolemy's theorem then says,

$$AC^2 = AB^2 + BC^2.$$

Leonardo da Vinci's proof. Start with right triangle AKE (fig. 8.7) and build the squares $EFGK$, $AKHI$, and $ABDE$ on the sides a and b and the hypotenuse c, respectively. Triangle BCD is a copy of the original triangle but turned through 180°. We now have the hexagon $ABCDEK$, which is bisected by the broken line KC. Connect G and H, producing the hexagon $AEFGHI$, which is bisected by the broken line IF. (Note that triangles AKE and HKG are mirror images of each other about the line IF, and therefore points I, K, *and F are collinear*.)

We claim that quadrilaterals $KABC$ and $IAEF$ are congruent, and therefore have the same area. To show this, turn $KABC$ through 90° counterclockwise about point A. We have $\angle IAE = 90° + \alpha = \angle KAB$ and $\angle ABC = 90° + \beta = \angle AEF$, causing AK to turn into AI, AB into AE, and BC into EF, while

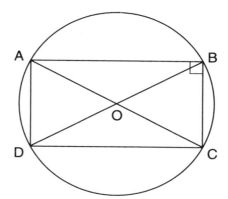

Figure 8.6. The Pythagorean theorem as a special case of Ptolemy's theorem

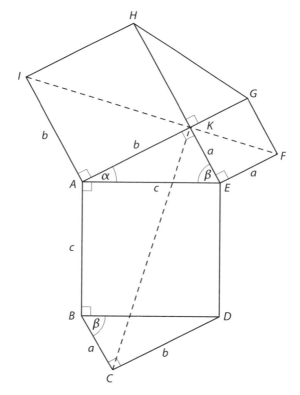

Figure 8.7. Leonardo da Vinci's proof

preserving the angles between these lines. Thus $KABC$ will overlap $IAEF$, making them equal in area. (Note that the two broken lines are exactly equal in length, despite the illusion that KC is longer than IF.)

It follows that hexagons $ABCDEK$ and $AEFGHI$ also equal areas. From the former hexagon subtract the congruent triangles AKE and DCB and from the latter hexagon, the congruent triangles AKE and HKG. Then $ABDE = AKHI + KEFG$; that is, $c^2 = b^2 + a^2$.

Loomis, on the authority of F. C. Boon, A.C. (*Miscellaneous Mathematics*, 1924), attributes this proof to Leonardo da Vinci (1452–1519).

James A. Garfield's proof. In figure 8.8, extend CB to D, with $BD = AC = b$. Draw $DE = CB = a$ and perpendicular to BD. Right triangles ACB and BDE are congruent, having two pairs of equal sides; consequently, angles ABC and EBD are complementary, and therefore ABE is a right angle. Now the area of trapezoid $ACDE$ is $(AC + ED) \times CD/2 = (b + a) \times (a + b)/2 = (a + b)^2/2$. This is also equal to the area of triangle ABE plus twice the area of triangle ACB, that is, $c^2/2 + 2(ab/2) = c^2/2 + ab$. Equating the two expressions, we get $a^2 + b^2 = c^2$ (as before, all line segments are nondirected). This proof has become famous because it was proposed by James A. Garfield (1831–1881), the future twentieth president of the United States. To quote Loomis, the demonstration "hit upon the General in a mathematical discussion with other M.C.'s [Members of Congress] about 1876."

Ann Condit's proof. Let the right triangle be ABC (fig. 8.9). Draw the squares $ACDE$, $BCFG$, and $ABHI$, and connect the corners D and F. Let CP be the bisector of the hypotenuse AB, and let its backward extension meet DF at R. We claim that $AP = PC$ and $PR \perp DF$.

To prove the first statement, we note that since $\angle ACB$ is a right angle, triangle ABC can be inscribed in a circle with center at P and diameter AB. Thus $AP = PC$, each being a radius.

To prove the second statement, we note that triangles ABC and DFC are congruent, having two pairs of equal sides and a right angle at C. Therefore, $\angle CDF = \angle BAC$. Call this angle α. By the first statement, triangle ACP is isosceles, so $\angle ACP$ is also α. Hence $\angle DCR = 90° - \alpha$, and thus $\angle CRD = 90°$, as claimed.

We are about ready to prove the main theorem. From P draw PM, PN, and PL to the midpoints of ED, FG, and HI, respectively; these lines are parallel to the sides AE, BG, and AI of the three squares. We now find the areas of triangles PFC, PDC, and PAI (all three are shaded in fig. 8.9) and compare them with the areas of the corresponding squares; to avoid repetition, PFC will denote the area of triangle PFC, $AEDC$ the area of square $AEDC$, and so on. We have $PFC = \frac{FC \times FN}{2}$, FC being the base and FN the altitude. But $FN = \frac{FG}{2} = \frac{FC}{2}$, so $PFC = \frac{FC^2}{4} = \frac{1}{4} BCFG$. In the same way, $PDC = \frac{1}{4} ACDE$ and $PAI = \frac{1}{4} ABHI$.

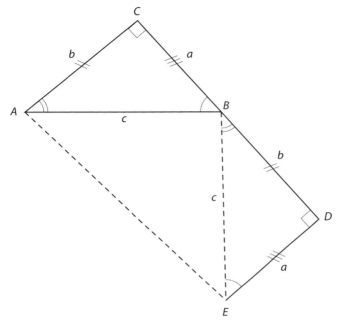

Figure 8.8. James A. Garfield's proof

We now use the fact that the areas of two triangles with the same base are to each other as their altitudes. Thus,

$$\frac{PDC + PFC}{PAI} = \frac{DR + RF}{AI} = \frac{DF}{AI} = \frac{AB}{AB} = 1.$$

Substituting the expressions we found earlier for *PDC*, *PFC*, and *PAI* and canceling the 1/4, we finally get

$$\frac{ACDE + BCFG}{ABHI} = 1,$$

that is, *ACDE + BCFG = ABHI*: the Pythagorean theorem.

This is a rather sophisticated proof, made all the more remarkable because it was proposed in 1938 by a student at Central Junior-Senior High School, South Bend, Indiana. Loomis, ever ready to praise a student, says of her: "This 16-year-old girl has done what no great mathematician, Indian, Greek, or modern, is ever reported to have done. It is the first ever [proof] devised in which all auxiliary lines and all triangles used originate at the middle point of the hypotenuse of the given triangle. It should be known as the Ann Condit Proof."

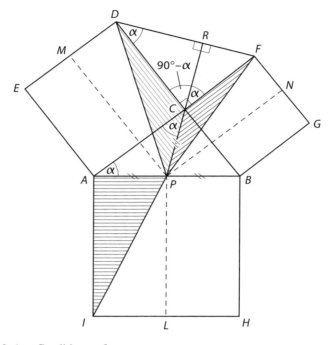

Figure 8.9. Ann Condit's proof

❖ ❖ ❖

The next two proofs are based on Propositions 35 and 36 of Book III of Euclid, which we give here in modern language:

Proposition 35.

  If through a point P inside a circle a chord is drawn meeting the circle at A and B, the product PA × PB is constant—it has the same value for all chords through P.

Proposition 36.

  If through a point P outside a circle a chord is drawn meeting the circle at A and B, the product PA × PB is equal to PT², where PT is the length of the tangent line from P to the circle. (As a consequence, PA × PB is constant for all chords through P.)

Let P be a point inside a circle with center O (fig. 8.10). Consider two chords through P, one passing through O and the other perpendicular to it. Let these chords intercept the circle at A, B and C, D, respectively. Now AB is the

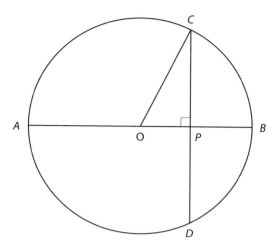

Figure 8.10. A little-known proof using a circle

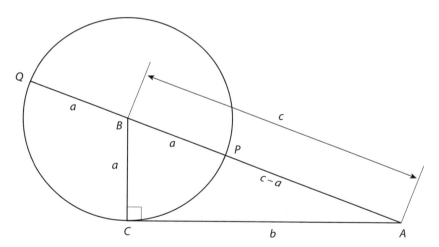

Figure 8.11. Another proof using a circle

perpendicular bisector of CD, so $\angle OPC = 90°$ and $PC = PD$. By Proposition 35 we have $PA \times PB = PC \times PD = PC^2$. But $PA = OA + OP$ and $PB = OB - OP = OA - OP$ (OA being the radius), so $(OA + OP) \times (OA - OP) = OA^2 - OP^2 = PC^2$. Replacing OA with OC, we finally get $OC^2 - OP^2 = PC^2$ or $OC^2 = OP^2 + PC^2$, the Pythagorean theorem for right triangle OPC.

Again, consider the right triangle ACB (fig. 8.11). With B as center, draw a circle with radius BC, touching AC at C. Let the extended line AB meet the

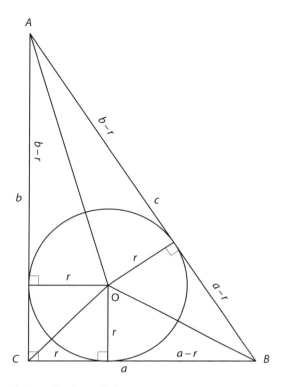

Figure 8.12. A third proof using a circle

circle at P and Q. By Proposition 36 we have $AP \times AQ = AC^2$; that is, $(c - a) \times (c + a) = b^2$, from which we get $a^2 + b^2 = c^2$ by expanding the left side.

We bring one more proof based on a circle, this one of an entirely different nature. In the right triangle ABC (fig. 8.12) inscribe a circle with center O and radius r, and draw the three radii perpendicular to the sides. This divides each side into two parts as shown. In particular, $c = (a - r) + (b - r) = a + b - 2r$, from which we get $r = \frac{a+b-c}{2}$. We now find the area of ABC in two ways: directly, as $A = \frac{ab}{2}$; and, by dividing ABC into the three triangles AOB, BOC, and COA, each with altitude r, as $A = \frac{ar}{2} + \frac{br}{2} + \frac{cr}{2} = \frac{(a+b+c)r}{2} = \frac{a+b+c}{2} \times \frac{a+b-c}{2} = \frac{(a+b)^2 - c^2}{4}$. Equating the two expressions and simplifying, we get $a^2 + b^2 = c^2$. Loomis gave several variants of this proof; and to secure his place in history, he added: "This solution was devised by the author Dec. 13, 1901, *before* [italics in the original] receiving Vol. VIII, 1901, p. 258, Am. Math. Mo. [the *American Mathematical Monthly*], where a like solution is given."

❖   ❖   ❖

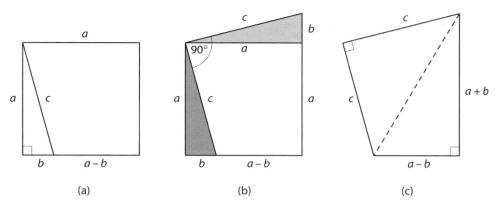

Figure 8.13. A proof based on rotation

New demonstrations of the Pythagorean theorem are being offered even at the time of writing. I refer the reader to Alexander Bogomolny's excellent Web site, "The Pythagorean Theorem and Its Many Proofs" (www.cut-the-knot.org /pythagoras/index.shtml), from which the following elegant proof is taken:

> Complete the right triangle (a, b, c) to a square with side a (fig. 8.13a). Turn the triangle counterclockwise through 90° about its top vertex (fig. 8.13b), and then delete the original triangle. This produces the quadrilateral shown in fig. 8.13c, whose area is obviously equal to the area of the square. Thus, $a^2 = \frac{c^2}{2} + \frac{(a-b)(a+b)}{2}$, from which we get, after simplifying, $a^2 + b^2 = c^2$.[5]

Here is a rather unusual proof, notable for its innovative approach even if its validity may rest on somewhat shaky grounds; I call it a "proof by differentials." Figure 8.14 shows one quadrant of a circle with center at O and radius a. Let $P(x, y)$ and $Q(x + dx, y + dy)$ be two neighboring points on this circle, where dx and dy denote "infinitesimally small" quantities. As P moves along the circle toward Q, the small triangle-like shape QRP (with its right angle at R) is nearly similar to triangle OSP, the similarity getting more precise the closer P is to Q. In the limit as $P \to Q$ we have $\triangle QRP \sim \triangle OSP$, from which we get

$$\frac{QR}{RP} = \frac{OS}{SP}$$

But $OS = x$, $SP = y$, $QR = -dy$, and $RP = dx$ (note that all line segments here are *directed*, hence the negative sign in front of dy). Thus, $\frac{-dy}{dx} = \frac{x}{y}$, from which we get, after cross-multiplying,

$$x\,dx + y\,dy = 0. \tag{4}$$

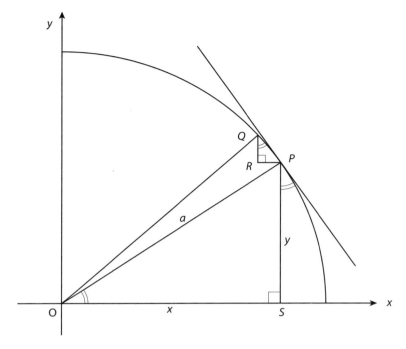

Figure 8.14. A proof by differentials

This is a differential equation whose general solution is

$$x^2 + y^2 = c, \tag{5}$$

where c is the constant of integration, as yet arbitrary. To determine it, we note that when $x = 0$ (P at the top of the circle), we have $y = a$. Putting this in equation (5), we find that $c = a^2$, so we get

$$x^2 + y^2 = a^2.$$

❖   ❖   ❖

To conclude this rich gallery of demonstrations, here (fig. 8.15) is a proof based on *tessellation*, filling the entire plane with a single pattern without leaving empty spaces and without overlapping. It is a "proof without words," so we present it without further explanation.[6]

In judging the relative merit of these proofs, we must first ask: What standard should be used? We expect a good proof to be as simple as possible, but what is "simple"? Is it the number of lines a proof takes up? The number of explanatory words? Perhaps a better criterion would be the number of earlier theorems on which a particular proof rests. The proof shown in figure 8.11

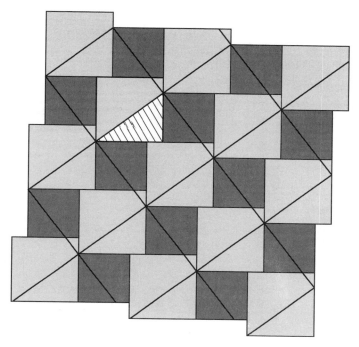

Figure 8.15. A proof based on tessellation

seems simple enough, but it depends on several properties of a circle, each of which is a theorem in its own right. By this standard, Euclid's proof I 47, the one that Schopenhauer derided as a "trap," is perhaps the simplest of all, as it relies on a bare minimum of earlier theorems. One might reflect whether, had Euclid known the four hundred or so proofs we know today, he would still have chosen the one he gave as I 47. The answer may well be yes.

Notes and Sources

1. Reston, Va.: National Council of Teachers of Mathematics, 1968. Henceforth this work will be referred to as Loomis.

2. *Original Investigation, or How to Attack an Exercise in Geometry* (Columbus, Ohio: Bonded Scale and Machine Company, 1952), with preface by Arthur Gluck. Other sources consulted are *The National Cyclopaedia of American Biography* (1916; rpt. Ann Arbor, Mich., 1967), vol. 15, p. 186; and David E. Kullman (Miami University), "Elisha S. Loomis, 1852–1940," 2004, on the Web: http://www.bgsn.edu/departments/math/Ohio=section/bicen/esloomis.html.

3. Kullman, "Elisha S. Loomis," p. 2.

4. Readers familiar with trigonometry will recognize in Ptolemy's theorem the

addition formula $\sin(\alpha + \beta) = \sin\alpha\cos\beta + \cos\alpha\sin\beta$. For a more detailed discussion of Ptolemy's theorem, as well as a biographical sketch of Ptolemy, see *Trigonometric Delights*, pp. 24–25 and 91–94.

5. Bogomolny credits this proof to W. J. Dobbs, *Mathematical Gazette*, 7 (1913–1914), p. 168.

6. I found this proof in Roger B. Nelsen, *Proofs without Words II: More Exercises in Visual Thinking* (Washington, D.C.: Mathematical Association of America, 2000), p. 3. Nelsen attributes it to Annairizi of Arabia (ca. 900 CE).

The Folding Bag

Simplicity, simplicity, simplicity. I say, let your
affairs be as two or three, and not a hundred or a
thousand; instead of a million count half a dozen,
and keep your accounts on your thumb nail.
—Henry David Thoreau, *Walden* (1854)

Here is what I believe to be the shortest—and perhaps the most
elegant—proof of the Pythagorean theorem. But first, two preliminary
notes.

1. As already noted in chapter 3, the theorem is valid not just for the
squares built on the three sides of a right triangle, but for *any* shapes—
as long as they are similar. In particular, we may choose an arbitrary
polygon as our representative shape. Since the areas of similar polygons
are in the same ratio to each other as the squares of their corresponding
sides, it is enough to prove the theorem for this particular polygon.

2. The phrase "built on" has usually been interpreted to mean that the
squares—or whatever similar shapes we choose instead—are to be built
on the *outside* of the right triangle. But nowhere is this actually man-
dated! We are free, in fact, to build any one of the three shapes—or all
three of them—on the *inside* of the given triangle.

And now for the proof. Which polygon shall we use? The simplest
choice would be a triangle; in fact, why not the *given triangle*? Referring to figure S4.1, we saw on page 41 that right triangles *ACB*, *ADC*,
and *CDB* are similar. And since the last two triangles dissect the first, we
have $S_{ACB} = S_{ADC} + S_{CDB}$, where S stands for area. This is precisely the
generalized form of the Pythagorean theorem, as stated in *Euclid* VI 31
(see p. 41)—QED.

I must confess I have a personal stake in this proof. Some years ago I arrived at it in a flash. But I first had to overcome traditional thinking: I had
to force myself to look at shapes other than squares, and to think of them as
built on the inside rather than outside of the triangle. Once these psycho-
logical barriers were overcome, the rest fell into place immediately. Visions

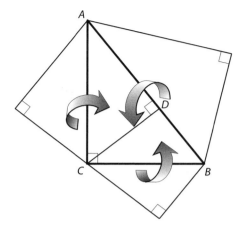

Figure S4.1. The Folding Bag proof

of immortal fame momentarily flashed through my mind, but I knew better: after two thousand years and some four hundred proofs, the chances of coming up with a new proof—and a short one at that—are almost nil. Still, I looked up Loomis's book, and sure enough, "my" proof appeared there as geometric proof number 230, with the comment that it is an original proof proposed on June 4, 1934, by Stanley Jashemski, age nineteen, of Youngstown, Ohio, "a young man of superior intellect." Oh well, someone much younger than I had beat me by seventy years![1] As a small consolation, I would like to propose a name for this proof: the *Folding Bag*, which is what came to my mind when I looked at the three triangles as they neatly fold up into the original triangle.

Notes and Sources

1. However, Loomis brings a second proof, essentially equivalent to this one but listed as algebraic proof number 96, to which he adds the following comment: "Devised by the author, July 1, 1901, and afterwards, Jan. 13, 1934, found in Fourrey's *Curio Geom.*, p. 91, where credited to R. P. Lamy, 1685." In light of this, one must wonder if there is anything new under the sun at all. See Elisha Scott Loomis, *The Pythagorean Proposition*, (Washington, D.C.: National Council of Teachers of Mathematics, 1968), pp. 85 and 230–231.

Einstein Meets Pythagoras

$$E = m(a^2 + b^2) = mc^2.$$
—An anonymous parody

At the age of twelve, Albert Einstein (1879–1955) received a little geometry book that he immediately devoured with great pleasure, affectionately calling it his "holy geometry booklet." As he wrote in his *Autobiographical Notes*, "Here were assertions, as for example the intersection of the three altitudes of a triangle at one point, that—though by no means evident—could nevertheless be proved with such certainty that any doubt appeared to be out of the question. This lucidity and certainty made an indescribable impression upon me."[1] He went on to say:

> An uncle told me about the Pythagorean theorem before the holy geometry booklet had come into my hands. After much effort I succeeded in "proving" this theorem on the basis of the similarity of triangles; in doing so it seemed to me "evident" that the relations [ratios] of the sides of the right-angled triangles would have to be completely determined by one of the acute angles. Only whatever did not in similar fashion seem to be "evident" appeared to me to be in need of any proof at all."

Einstein's "proof" (he was careful to put the word in quotation marks, evidently not wishing to take credit for it) was reconstructed by his biographer and collaborator Banesh Hoffmann;[2] it turns out to be identical with the first of the "algebraic proofs" in Elisha Scott Loomis's book (attributed there to Legendre but actually being Euclid's second proof; see p. 41). Einstein's early enchantment with the Pythagorean theorem was to bear fruit ten years later: it would play a key role, first in its four-dimensional form, in his special theory of relativity; and later in a vastly expanded form, in general relativity.

Notes and Sources

1. Trans. and ed. Paul Arthur Schlipp (La Salle, Ill.: Open Court, 1979), pp. 9–11.

2. *Albert Einstein: Historical and Cultural Perspectives*, ed. Gerald Holton and Yehuda Elkana (Princeton, N.J.: Princeton University Press, 1982), pp. 92–93.

A Most Unusual Proof

Please forget everything you have learned
in school; for you haven't learned it.
—Edmund Landau,
Foundations of Analysis (1960), p. v.

"There are no trigonometric proofs [of the Pythagorean theorem], because all the fundamental formulas of trigonometry are themselves based upon the truth of the Pythagorean theorem. Trigonometry *is* because the Pythagorean Theorem *is*." Thus declared Elisha Loomis toward the end of his book, *The Pythagorean Proposition*. Indeed, if in the right triangle ABC (fig. S6.1) we define the sine and cosine of the angle α as $\sin \alpha = a/c$, $\cos \alpha = b/c$, then $\sin^2 \alpha + \cos^2 \alpha = (a/c)^2 + (b/c)^2 = (a^2 + b^2)/c^2 = c^2/c^2 = 1$; and having used the Pythagorean theorem in proving this identity, we cannot use the same identity to prove the Pythagorean theorem, or else we would be committing the worst of mathematical sins—using a circular argument. So when proving the Pythagorean theorem, trigonometry is off-limits.

Or is it? At about the same time that Loomis worked on the second edition of his book, Edmund Landau in Germany published a textbook, *Differential and Integral Calculus*,[1] that became a model of rigorous exposition. Landau (1877–1938) was professor of mathematics at Göttingen University, the world-renowned center of mathematical research until World War II. He is best known for his work in analytic number theory—the application of analytic methods (i.e., the calculus) to the study of integers. He published over 250 papers and wrote several seminal works in his field, among them *Handbook of the Theory and Distribution of the Prime Numbers* (in two volumes, 1909) and *Lectures on Number Theory* (in three volumes, 1927).

Landau was famous for his uncompromising pedantry and rigor. In his teaching he shunned all reference to geometry, calling it "Schmieröl" (grease, a slur in German). His 372-page calculus book—a far cry from today's 1,000-plus page texts—contains not a single illustration. In a

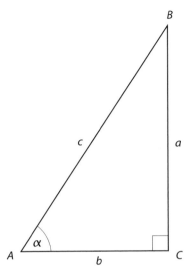

Figure S6.1. $\sin \alpha = a/c$, $\cos \alpha = b/c$

terse, definition-theorem-proof style, supplemented here and there by a few examples, he developed the calculus from first principles up to its highest levels. Of special interest to us here is chapter 16 on the trigonometric functions. It begins thus:[2]

Theorem 248:

$$\sum_{m=0}^{\infty} \frac{(-1)^m}{(2m+1)!} x^{2m+1}$$

converges everywhere.

(This, of course, is the power series $x - x^3/3! + x^5/5! - + \ldots$.) This is followed by

Definition 59:

$$\sin x = \sum_{m=0}^{\infty} \frac{(-1)^m}{(2m+1)!} x^{2m+1}$$

sin is to be read "sine."

Next comes the definition of $\cos x$ as the power series $1 - x^2/2! + x^4/4! - + \ldots$, followed by several theorems establishing the familiar properties of these functions, including the addition formula $\cos(x+y)$

$= \cos x \cos y - \sin x \sin y$ and the even-odd relations $\sin(-x) = -\sin x$ and $\cos(-x) = \cos x$. Then,

Theorem 258:

$$\sin^2 x + \cos^2 x = 1.$$

Proof:

$$1 = \cos 0 = \cos(x - x) = \cos x \cos(-x) - \sin x \sin(-x)$$
$$= \cos^2 x + \sin^2 x.$$

Thus, out of the blue and without any mention by name, the most famous theorem in mathematics is introduced: the Pythagorean theorem.

No doubt most of us would regard such an approach as pedantic sophistry. Essentially, it turns the tables around: the trigonometric functions are defined through their infinite series, are given the names "sine" and "cosine," and are then treated in a strictly formal manner, without the slightest recognition of their role in geometry, let alone right triangles. Of course, the implied assumption is that the reader will see these functions for what they are, and conclude that what looks like a duck and walks like a duck, is in fact a duck. But then again, perhaps Landau did *not* want us to make such an assumption.

To the skeptic who may wonder how two infinite series, each a function of x and each being squared, will add up to the constant 1, here is a demonstration (though by no means a proof). Let us take just the first two terms of each series, square them, and add:

$$(1 - x^2/2!)^2 + (x - x^3/3!)^2$$
$$= 1 - 2x^2/2! + x^4/(2!)^2 + x^2 - 2x^4/3! + x^6/(3!)^2.$$

The second and fourth terms cancel out, leaving us with $1 - x^4/12 + x^6/36$. If we repeat the calculation with the first *three* terms of each series—a tedious but otherwise straightforward task—we will find that the x^2 *and* x^4 terms cancel out. This pattern continues: the more terms we take, the more powers of x cancel out, leaving us with the leading 1 and "leftover" terms with increasingly large denominators. As the number of terms in each series tends to infinity, the sum of their squares tends to 1.

So, is this an acceptable proof of the Pythagorean theorem? It depends, of course, to whom the proof is addressed. No doubt most of us would prefer a traditional proof, using a right triangle or some other geometric figure. But to Landau this was unacceptable; he based all his arguments on first principles, in this case infinite series. Whatever significance these series may have in the "real" world did not matter to him in the least. He was the embodiment of the pure mathematician *par excellence*.

Notes and Sources

1. Trans. Melvin Hausner and Martin Davis (New York: Chelsea, 1965).

2. The subsequent material is adapted from *Trigonometric Delights*, pp. 192–197. Therein the reader will find a biographical sketch of Landau.

A Theme and Variations

Pythagoras was the first to divine a great truth
about right triangles. He pointed out that the sum
of the squares of the sides is equal to the square of
the hypotenuse, a formula that is hammered into
every teenage brain that wanders into a geometry
classroom from Des Moines to Ulan Bator.
—Leon Lederman (with Dick Teresi),
   in *The God Particle: If the Universe Is the Answer,
   What Is the Question?*, p. 66

OK, so $a^2 + b^2 = c^2$, and we have some four hundred ways to prove it. So what else is there to say?

A lot. For some inexplicable reason, no theorem has generated more commentaries, variations, applications, and curiosa than the Pythagorean theorem. Some of these border on the trivial, whereas others are quite profound. Following is a sample, culled from various sources; no doubt many more can be found.

As already mentioned, the generalized version of the theorem as stated in *Euclid* VI 31 (see p. 41) allows us to replace the squares on the sides of a right triangle by arbitrary shapes—polygonal or not—provided they are similar.[1] Figure 9.1 shows the theorem for regular pentagons, and figure 9.2, for semicircles built on the sides as diameters.

Of course, the same is true for *full* circles as well, although in this case the three circles will partially overlap and thus clutter the picture a bit (fig. 9.3). This leads to an interesting result. We know that through any three noncollinear points (points not on one line), one and only one circle passes; put differently, any triangle can be circumscribed by a unique circle, called the *circumcircle* of the triangle. Now suppose the vertices form a *right* triangle. A well-known theorem (*Euclid* III, 31) says that the right angle subtends the diameter of this circumcircle; in other words, the diameter coincides with the hypotenuse. Consequently, *in any right triangle, the sum of the areas of the circles built on the two legs is equal to the area of the circumcircle.*

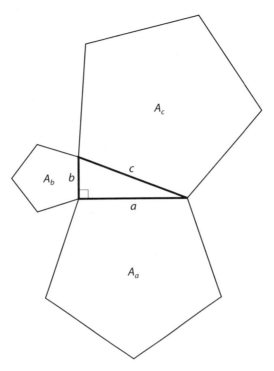

Figure 9.1. The Pythagorean theorem applied to regular pentagons: $A_c = A_a + A_b$

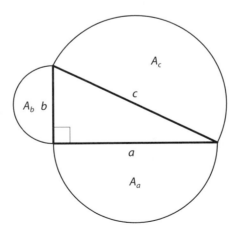

Figure 9.2. The Pythagorean theorem applied to semicircles: $A_c = A_a + A_b$

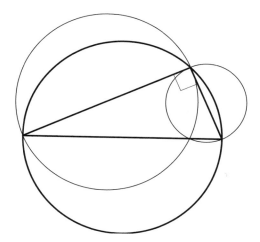

Figure 9.3. The Pythagorean theorem applied to full circles

A somewhat more interesting result involves the *lune of Hippocrates* (named after Hippocrates of Chios, fl. 460 BCE). Consider the quadrant *AOB* of a circle with center at *O* and radius *OA* = *OB* (fig. 9.4). Draw a semicircle with *AB* as diameter. The lune of Hippocrates is the crescent-shaped region bounded by the semicircle and the arc *AB* of the original circle. Surprisingly, this lune has the same area as triangle *AOB*. The proof is rather simple:

> Area of lune = area of semicircle on *AB* minus area of circular segment *AB*.

The area of the semicircle on *AB* is $\frac{1}{2} \times \frac{\pi AB^2}{4}$, and the area of the circular segment *AB* is the difference between the area of the quarter circle with radius *OA* and triangle *AOB*, that is, $\frac{1}{4} \times \pi OA^2 - \frac{OA^2}{2}$. We thus have

$$\text{Area of lune} = \frac{1}{2} \times \frac{\pi AB^2}{4} - \left(\frac{1}{4} \times \pi OA^2 - \frac{OA^2}{2} \right).$$

But $AB^2 = OA^2 + OB^2 = 2OA^2$. Putting this back in the expression above, we get $\frac{\pi \times 2OA^2}{8} - \frac{\pi \times OA^2}{4} + \frac{OA^2}{2}$. The first two terms cancel out, so

$$\text{Area of lune} = \frac{OA^2}{2} = \text{Area of } \Delta AOB.$$

Surprisingly, the result is independent of π, even though the lune was generated from the arcs of two circles. Moreover, since a triangle can always be "squared" (in the sense that we can construct, using a straightedge and compass, a square

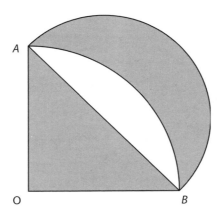

Figure 9.4. Lune of Hippocrates

equal in area to the triangle), the result shows that this particular lune can be squared. By contrast, a *full* circle, as is well known, cannot be squared.[2]

A famous configuration based on the Pythagorean theorem allows us to construct the square root of any integer n, provided we have already constructed the square root of $n - 1$. In figure 9.5, let OP_1 be the line segment from 0 to 1 along the number line. At P_1 erect $P_1P_2 \perp OP_1$ and of length 1. The radius vector from O to P_2 has length $r_2 = \sqrt{1^2 + 1^2} = \sqrt{2}$. At P_2 erect $P_2P_3 \perp OP_2$ and of length 1. The radius vector from O to P_3 has length $r_3 = \sqrt{(\sqrt{2})^2 + 1^2} = \sqrt{2 + 1} = \sqrt{3}$. Continuing in this manner n times, we get $r_n = \sqrt{n}$. The points $P_1, P_2, \ldots$, form the spiral-like configuration shown in the figure.

Let a circle be inscribed in a right triangle whose three sides have integer lengths (and thus form a Pythagorean triple). Then the radius r of this *incircle* is also an integer. To prove this, we go back to fig. 8.12 (p. 110), from which we derived the formula $r = \frac{a+b-c}{2}$. To show that this formula always produces an integer, we must show that $a + b - c$ is an *even* integer. This follows from the equation $a^2 + b^2 = c^2$; for, if a and b are both even, so is the sum of their squares, which is c^2. But then c must also be even, because the square of an odd number is always odd. On the other hand, if a is even and b odd, the

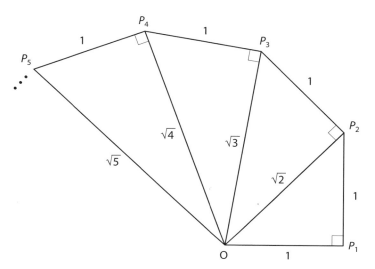

Figure 9.5. The square-root spiral

sum of their squares is odd, and therefore c is also odd. In either case, the expression $a + b - c$ will be even, as claimed.[3]

❖ ❖ ❖

In the right triangle ABC (fig. 9.6), let the perpendicular CD from the right angle to the hypotenuse have length d. We then have

$$\frac{1}{a^2} + \frac{1}{b^2} = \frac{1}{d^2}.$$

To prove this, write $1/a^2 + 1/b^2 = (a^2 + b^2)/a^2b^2 = c^2/a^2b^2$. Now the area of $\triangle ABC$ can be written in two ways, either as $ab/2$ or as $cd/2$, so we have $ab = cd$, from which $c = ab/d$. Putting this back in the expression c^2/a^2b^2 and simplifying, we get the stated result. I like to call this "the Little Pythagorean theorem"; we will have a chance to use it in chapter 10 in a rather unusual way.

❖ ❖ ❖

The Pythagorean theorem can easily be generalized to *any* triangle, not just a right one, but at a price: to the right-hand side of the equation $c^2 = a^2 + b^2$ we must add a "correction term," $-2ab \cos C$, where C is the angle opposite side c. This, of course, is the Law of Cosines, familiar to us from solving the side-angle-side (*SAS*) case in trigonometry. But the nontrigonometric form of this law was already known to Euclid; in fact, it appears in Book II of the *Elements* as Propositions 12 and 13, which we combine here into a single statement using modern language:

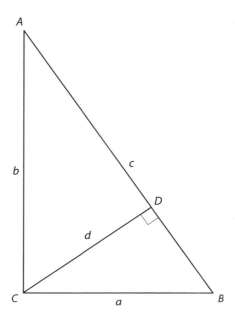

Figure 9.6. $\dfrac{1}{d^2} = \dfrac{1}{a^2} + \dfrac{1}{b^2}$

In any triangle, the square of a side lying opposite an obtuse (acute) angle is equal to the sum of the squares of the adjacent sides, plus (minus) twice the product of either of the adjacent sides and the perpendicular projection of that side on the other.[4]

Figure 9.7a shows a triangle with sides a, b, and c, where c lies opposite the obtuse angle C. Drop the perpendicular from A to the extension of BC, and call the sides of the small right triangle so formed x and y. We have

$$c^2 = (a + x)^2 + y^2 = (a^2 + 2ax + x^2) + y^2$$
$$= (x^2 + y^2) + a^2 + 2ax$$
$$= b^2 + a^2 + 2ax.$$

A similar derivation holds when C is an acute angle (fig. 9.7b), except that we now have $c^2 = (a - x)^2 + y^2$. We note that in the former case, $x = b \cos(180° - C) = -b \cos C$, while in the latter case $x = b \cos C$. We can therefore combine the two cases into one equation, $c^2 = a^2 + b^2 - 2ab \cos C$, where $\cos C$ is positive or negative depending on whether C is acute or obtuse. The Law of Cosines thus obviates the need to distinguish between the two cases.

❖   ❖   ❖

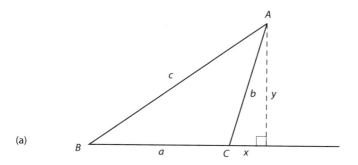

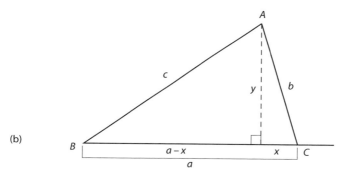

Figure 9.7. The Law of Cosines: (a) the obtuse case; (b) the acute case

Consider again any triangle ABC, construct the squares on all three sides, and connect their corners externally (fig. 9.8). Denote by x, y, and z the lengths of the connecting line segments. We then have the following elegant result:

$$x^2 + y^2 + z^2 = 3(a^2 + b^2 + c^2).$$

To prove this, denote the angles of the outer triangles at A, B, and C by α, β, and γ. Applying the Law of Cosines to the three exterior triangles, we have

$$x^2 = b^2 + c^2 - 2\,bc \cos \alpha,$$
$$y^2 = c^2 + a^2 - 2\,ca \cos \beta,$$
$$z^2 = a^2 + b^2 - 2\,ab \cos \gamma.$$

Adding the three equations, we get

$$x^2 + y^2 + z^2 = 2(a^2 + b^2 + c^2) - 2(bc \cos \alpha + ca \cos \beta + ab \cos \gamma). \quad (1)$$

Now in the central triangle ABC we have

$$a^2 = b^2 + c^2 - 2\,bc \cos A,$$
$$b^2 = c^2 + a^2 - 2\,ca \cos B,$$
$$c^2 = a^2 + b^2 - 2\,ab \cos C,$$

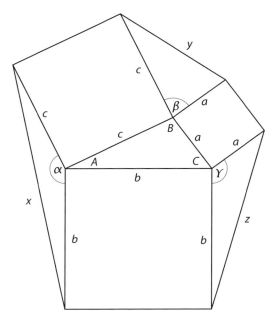

Figure 9.8. $x^2 + y^2 + z^2 = 3(a^2 + b^2 + c^2)$

where A, B, and C are the interior angles at the corresponding vertices. Adding, we get

$$a^2 + b^2 + c^2 = 2(a^2 + b^2 + c^2) - 2(bc \cos A + ca \cos B + ab \cos C), \quad (2)$$

and therefore

$$2(bc \cos A + ca \cos B + ab \cos C) = a^2 + b^2 + c^2. \quad (3)$$

Now, since the squares built on the sides a, b, and c form right angles at A, B, and C, we have $\alpha = 180° - A$, $\beta = 180° - B$, $\gamma = 180° - C$. Putting this back in equation (1) and using the identity $\cos(180° - \theta) = -\cos \theta$, we get

$$x^2 + y^2 + z^2 = 2(a^2 + b^2 + c^2) + 2(bc \cos A + ca \cos B + ab \cos C).$$

But the last term, in view of equation (3), is equal to $a^2 + b^2 + c^2$, so

$$x^2 + y^2 + z^2 = 2(a^2 + b^2 + c^2) + (a^2 + b^2 + c^2)$$
$$= 3(a^2 + b^2 + c^2).[5]$$

As an additional result, we note that each of the three external triangles has the same area as the inner triangle. For example, the area of the exterior triangle on the left is $\frac{1}{2} bc \sin(180° - A) = \frac{1}{2} bc \sin A$, which is the area of the inner triangle.

❖　❖　❖

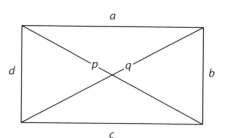

Figure 9.9. $p^2 + q^2 = a^2 + b^2 + c^2 + d^2$

Figure 9.9 shows a rectangle with sides a, b, c, and d and diagonals p and q. The diagonals divide the rectangle into two congruent right triangles, so we have $p^2 = a^2 + b^2$ and $q^2 = c^2 + d^2$. Adding the two equations, we get

$$p^2 + q^2 = a^2 + b^2 + c^2 + d^2,$$

which, in view of the fact that $a = c$, $b = d$, and $p = q$, is a rather trivial result. Now let us stretch the rectangle into a parallelogram (fig. 9.10). We still have $a = c$ and $b = d$, but the two diagonals now have different lengths, and, further-more, they divide the parallelogram into two scalene rather than right triangles, so the Pythagorean theorem does not apply to them. Nevertheless, the relation $p^2 + q^2 = a^2 + b^2 + c^2 + d^2$ still holds. To show this, denote the included angle between sides a and b by θ. We have $p^2 = a^2 + b^2 - 2\,ab\,\cos\theta$ and $q^2 = a^2 + d^2 - 2\,ad\,\cos(180° - \theta) = a^2 + d^2 + 2\,ad\,\cos\theta$. Replacing a^2 in the second equation with c^2 and adding the two equations, we get the required result. (This is no longer true, however, for an arbitrary quadrilateral.)

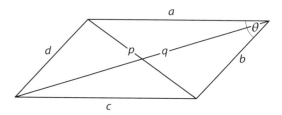

Figure 9.10. $p^2 + q^2 = a^2 + b^2 + c^2 + d^2$

A double application of the Pythagorean theorem leads to a famous for-mula for the area A of any triangle. Let the triangle have sides a, b, and c, and let $s = (a + b + c)/2$ be its *semiperimeter*. Then

$$A = \sqrt{s(s - a)(s - b)(s - c)}.$$

This formula is usually credited to the Greek mathematician, surveyor and engineer Heron, who lived sometime between 100 BCE and 100 CE (his name is also spelled Hero); however, according to the Arabic astronomer Al-Biruni (ca. 973–1048), it was actually discovered by Archimedes.[6] The remarkable thing about Heron's formula is that it allows us to calculate the area of a triangle solely from its three sides; that is to say, a, b, and c determine A uniquely. This is due to the fact that among all polygons, the triangle is the only *rigid* polygon: if a triangle can be constructed from three given line segments, it is unique. (This is not so for any other polygon; for example, from four equal sides we can construct infinitely many rhombi, each with a different area.) It is believed that Heron used this formula to calculate the area of tracts of land along the banks of the Nile after they had been inundated by the annual floods, when only traces of their original outline were left visible; this information was then used to give the owners of the lost land a tax break.

To prove Heron's formula, we refer to figure 9.11. From the top vertex we drop the altitude h to side a, dividing it into parts m and n. We have

$$m^2 + h^2 = b^2, \quad n^2 + h^2 = c^2.$$

Subtracting the second equation from the first, we get

$$m^2 - n^2 = b^2 - c^2.$$

But $m^2 - n^2 = (m + n)(m - n) = a(m - n)$, so $m - n = (b^2 - c^2)/a$. Adding this to the equation $m + n = a$ and solving for m and n, we get

$$m = \frac{a^2 + b^2 - c^2}{2a}, \quad n = \frac{a^2 - b^2 + c^2}{2a}.$$

Now,

$$
\begin{aligned}
h^2 = b^2 - m^2 &= b^2 - \frac{(a^2 + b^2 - c^2)^2}{4a^2} \\
&= \left(b + \frac{a^2 + b^2 - c^2}{2a} \right)\left(b - \frac{a^2 + b^2 - c^2}{2a} \right) \\
&= \frac{(2ab + a^2 + b^2 - c^2)(2ab - a^2 - b^2 + c^2)}{4a^2} \\
&= \frac{[(a+b)^2 - c^2][c^2 - (a-b)^2]}{4a^2} \\
&= \frac{(a+b+c)(a+b-c)(c+a-b)(c-a+b)}{4a^2}
\end{aligned}
$$

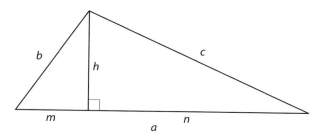

Figure 9.11. To prove Heron's formula

Each of the factors in the last equation can be written in terms of the semiperimeter $s = (a + b + c)/2$:

$$h^2 = \frac{2s \times 2(s - c) \times 2(s - b) \times 2(s - a)}{4a^2}$$
$$= \frac{4s\,(s - a)(s - b)(s - c)}{a^2}.$$

Thus,

$$h = \frac{2\sqrt{s(s - a)(s - b)(s - c)}}{a}.$$

We are finally ready to find the area of the triangle:[7]

$$A = \frac{ah}{2} = \sqrt{s(s - a)(s - b)(s - c)}.$$

As an extra bonus from Heron's formula, we give here without proof the formulas for the inradius r and exradius R of the inscribed and circumscribing circles of a triangle, respectively:

$$r = \frac{\sqrt{s(s - a)(s - b)(s - c)}}{s} = \frac{A}{s}, \quad R = \frac{abc}{4\sqrt{s(s - a)(s - b)(s - c)}} = \frac{abc}{4A}.$$

The Pythagorean theorem, of course, plays a key role in analytic, or coordinate geometry, invented by René Descartes (1596–1650) in 1637. The familiar distance formula, $d = \sqrt{(x_2 - x_1)^2 + (y_2 - y_1)^2}$, is well known to every algebra and calculus student (to the chagrin of those who take the liberty of "simplifying" it to $(x_2 - x_1) + (y_2 - y_1)$ and then argue with their professor over being

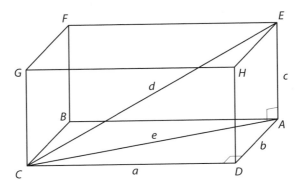

Figure 9.12. $d^2 = a^2 + b^2 + c^2$

marked wrong). Strangely, this formula did not appear in print until 1731, when the French mathematician Alexis Claude Clairaut (1713–1765) published his *Recherches sur les courbes à double courbure* ("Researches on curves of double curvature"), in which it appears as $\sqrt{\overline{x \mp a}^2 + \overline{y \mp b}^2}$ (note the archaic upper bars, a precursor of our modern parentheses).[8]

But there is no reason why we should restrict ourselves to two dimensions. The three-dimensional version of the Pythagorean theorem says: *In a rectangular box, the square of the space diagonal is equal to the sum of the squares of the three sides.*[9] Let the box have dimensions $a \times b \times c$ (fig. 9.12). Draw the base diagonal AC, and call its length e. In the horizontal triangle CDA we have $e^2 = a^2 + b^2$, and in the vertical triangle CAE we have $d^2 = e^2 + c^2$. Combining the two equations, we get

$$d^2 = a^2 + b^2 + c^2.$$

This equation also allows the formation of *Pythagorean quadruples*, integers (a, b, c, d) such that $d^2 = a^2 + b^2 + c^2$; examples are (3, 4, 12, 13) and (36, 77, 204, 221).

An even more interesting case arises if we require that all three *face diagonals* of the box have integer lengths. Denoting these diagonals by e, f, and g, we have $a^2 + b^2 = e^2$, $b^2 + c^2 = f^2$, and $c^2 + a^2 = g^2$. One solution, said to have been found by Euler, is $(a, b, e) = (240, 44, 244)$, $(b, c, f) = (44, 117, 125)$, and $(c, a, g) = (117, 240, 267)$, shown in figure 9.13. It is still an open question whether one can make the three face diagonals *and* the space diagonal d have integer lengths (which would add the equation $a^2 + b^2 + c^2 = d^2$ to the three equations given above). As of this writing, no such "Pythagorean cuboids" are known.[10]

❖   ❖   ❖

Plate 1. Postal stamps honoring Pythagoras and his contemporaries (courtesy Robin J. Wilson). *First row*: Greece, 1955; Greece, 1998. *Second row*: Macedonia, 1998; Nicaragua, 1971. *Third row*: Greece, 1994, showing Thales of Miletus; Sierra Leone, 1983, showing Raphael's *School of Athens*. *Fourth row*: Greece, 1961, showing Democritus.

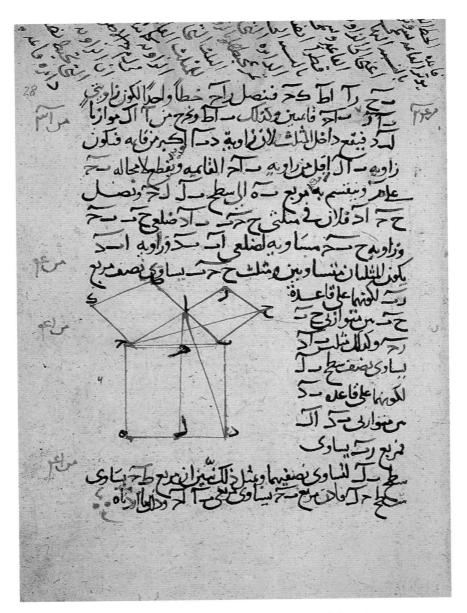

Plate 2. The Pythagorean theorem in an Arab text, from the eighth century

Plate 3. "The Cloud Gate" (popularly known as the "Bean"), by Anish Kapoor, at the Millennium Park, Chicago. Note how the rectangular grid of the pavement becomes a curved grid as seen on the reflective surface of the "Bean."

Plate 4. Aristarhos of Samos Airport

Plate 5. Pythagorio, Samos: the harbor

Plate 6. Pythagorio: Gymnasio Pythagoreio

Plate 7. Pythagorio: Pythagora Street

Plate 8. Pythagorio: statue of Pythagoras

ΑΡΙΣΤΑΡΧΟΣ Ο ΣΑΜΙΟΣ
320 – 250 π.Χ.
ARISTARCHOS OF SAMOS
320 – 250 B.C.
FIRST TO DISCOVER THE
EARTH REVOLTES AROUND
THE SUN.
COPERNIKUS COPIED ARISTARCHOS
1530 A.D.

Plate 9. Karlovasi, Samos: statue of Aristarchus

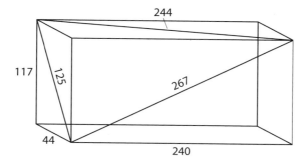

Figure 9.13. A "Pythagorean box"

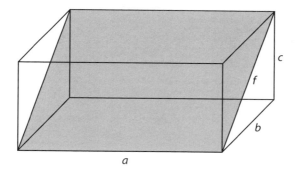

Figure 9.14. $A^2 = A_{base}^2 + A_{back}^2$

But there is more. The area of the shaded rectangle in figure 9.14 is $A = af = a\sqrt{b^2 + c^2}$, so $A^2 = a^2(b^2 + c^2) = a^2b^2 + a^2c^2$. But ab and ac are the areas of the base and back faces, respectively, so we have

$$A^2 = A_{base}^2 + A_{back}^2, \tag{4}$$

a Pythagorean-like equation relating the area of the front face of a right triangular prism to the areas of its sides. A still more interesting result arises if we connect pairs of nonadjacent vertices of the base, side, and back of the box, producing the sail-shaped triangle of figure 9.15. We then have the formula

$$A_{ACF}^2 = A_{ABC}^2 + A_{ABF}^2 + A_{CBF}^2. \tag{5}$$

This means that *the square of the front-face area of a right tetrahedron is equal to the sum of the squares of the other face areas*. The analogy of these formulas to the Pythagorean relations $c^2 = a^2 + b^2$ and $d^2 = a^2 + b^2 + c^2$ is quite striking.[11]

To prove equation (5), we refer to figure 9.16, in which we kept the same vertex letters as in figure 9.15. In right triangle ABC, drop the altitude p from B to AC. We saw on page 127 that $1/p^2 = 1/a^2 + 1/b^2$, from which we get

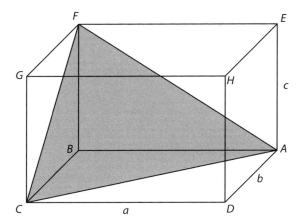

Figure 9.15. $A_{ACF}^2 = A_{ABC}^2 + A_{ABF}^2 + A_{CBF}^2$

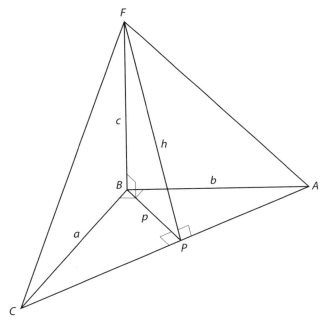

Figure 9.16. A right tetrahedron

$p = ab / \sqrt{a^2 + b^2}$. In triangle ACF, drop the altitude h from F to AC; this altitude is also the hypotenuse of right triangle PBF, so we have

$$h = \sqrt{p^2 + c^2} = \sqrt{\frac{a^2 b^2}{a^2 + b^2} + c^2} = \sqrt{\frac{a^2 b^2 + b^2 c^2 + c^2 a^2}{a^2 + b^2}}.$$

Thus the area of triangle ACF is

$$A_{ACF} = \frac{1}{2} base \times height = \frac{1}{2}\sqrt{a^2 + b^2} \cdot \sqrt{\frac{a^2 b^2 + b^2 c^2 + c^2 a^2}{a^2 + b^2}}$$

$$= \sqrt{\frac{a^2 b^2 + b^2 c^2 + c^2 a^2}{4}}.$$

Squaring, we get

$$A_{ACF}^2 = \frac{a^2 b^2 + b^2 c^2 + c^2 a^2}{4} = \left(\frac{ab}{2}\right)^2 + \left(\frac{bc}{2}\right)^2 + \left(\frac{ca}{2}\right)^2$$

$$= A_{ABC}^2 + A_{ABF}^2 + A_{CBF}^2.$$

But why stop at three dimensions? Mathematicians routinely work with spaces of four, five, or any number of dimensions, even if we cannot see them in a literal sense. For example, a four-dimensional sphere of radius r and center at the origin "looks" like this: $x^2 + y^2 + z^2 + t^2 = r^2$ (don't ask me to draw it, though). And once we have broken out of the confinement of our familiar, three-dimensional world, there is no reason why we cannot extend the number of dimensions to infinity. It is strange indeed to think of a "rectangular box" of dimensions $x_1, x_2, \ldots$. But the crux of the matter is that in contrast to the physical sciences, mathematics offers its practitioners complete latitude to create their own world, subject only to the laws of logic that require it to be free of internal contradiction. So there is nothing that prevents us from thinking of, and even working in, an infinitely many dimensional space.

However, if our infinitely many dimensional box is to preserve any semblance to an ordinary box, we ought to be able to find the length of its space diagonal. Assuming the Pythagorean theorem to be valid in this space, the length of the diagonal is given by the formula

$$d = \sqrt{x_1^2 + x_2^2 + x_3^2 + \cdots}.$$

Of course, for this formula to make any sense at all, the sum of squares under the radical sign must have a finite value—it must *converge*. This at once thrusts the Pythagorean theorem from its original geometric setting into the domain of analysis, the branch of mathematics that deals with continuity, change, and the limit process.

All this may sound like the height of abstract thinking, but in fact it has found many important, down-to-earth applications. I will discuss here one of them, whose roots go back to the very same subject that inspired Pythagoras to shape his views of the universe: acoustics, the science of sound.

Every sound is a mixture of many different vibrations, each represented by a simple sine wave with its own amplitude and frequency. A *musical* sound is made up of sine waves whose frequencies are integral multiples of a lowest, or fundamental, frequency. The fundamental frequency determines the pitch of the sound—its position on the musical staff—whereas the higher frequencies, known as harmonics or overtones, determine its timbre, the quality that makes a violin sound different from a clarinet, even when they play the same note. Let the fundamental frequency be f, measured in cycles per second (the note A above middle C, for example, has a frequency of 440 cps). Its overtones have the frequencies $2f$, $3f$, ... In addition, each overtone has its own amplitude a_n, so we can express its vibration as $a_n \sin(2\pi nf) t$, $n = 1, 2, 3, \ldots$, where t denotes time. The actual sound is the sum of these vibrations, that is, $\sum_1^\infty a_n \sin(2\pi nf)t$.[12]

Now it is shown in the theory of sound that the energy of a pure sine wave is proportional to the square of its amplitude. Therefore, the total energy carried by the sound is given by the expression $a_1^2 + a_2^2 + a_3^2 + \ldots$ —the very same expression whose square root represents the length of the diagonal of an infinitely many dimensional box. That this sum does indeed converge is a consequence of the fact that the total energy content of the sound cannot exceed the initial energy put into it when the sound was generated (for example, by plucking a string), and this energy, by necessity, is finite.

So we have come, in a way, full circle: a purely abstract mathematical idea finding its way back into music. How would Pythagoras have reacted to the idea of an infinitely many dimensional space? If the historical record is any indication, he would have been horrified. The Greeks had a deeply rooted suspicion of infinity, effectively banishing it from their mathematics. So it is unlikely that Pythagoras would have appreciated the leap from our comfortable, three-dimensional world to one of infinitely many dimensions.

But who knows? Perhaps against all odds, he might have been thrilled by this unexpected connection between music and mathematics. We can only speculate.

Notes and Sources

1. This follows from the fact that the areas of similar polygons are to one another as the squares of their corresponding sides. Thus, if the sides of one polygon are t times as long as the corresponding sides of the other, the first polygon will have an area t^2 times

as large as the second. Multiplying both sides of the equation $c^2 = a^2 + b^2$ by t^2 proves the stated result.

2. More on lunes can be found in Tobias Dantzig, *The Bequest of the Greeks* (New York: Charles Scribner's Sons, 1955), chap. 10.

3. It is proved in elementary number theory that a and b cannot both be odd. See Appendix B.

4. For additional commentary on the two propositions, see *Euclid: The Elements*, trans., with introduction and commentary, by Sir Thomas Heath (in 3 vols.; New York: Dover, 1956), vol. 1, pp. 403–409.

5. I found this beautiful result in J. L. Heilbron, *Geometry Civilized: History, Culture, and Technique* (Oxford, U.K.: Clarendon Press, 1998), p. 164. This was shortly before I was due for a relatively minor surgery. Knowing I would have to spend many hours at the hospital waiting for the operation, and many more hours recovering from it, I scribbled down on a sheet of paper a few theorems I wanted to prove to myself, including this one. Alas, the nurse in charge of the presurgery room would not allow anything to be carried in, not even my sheet of paper. So I quickly memorized the theorem, and while awaiting for the anesthetist to arrive, I managed to prove it mentally just before becoming unconscious. If nothing else, at least it helped me overcome the anxiety one always has before surgery—proof (if one is needed) that mathematics can do some good, on occasion.

6. See Bartel L. van der Waerden, *Science Awakening: Egyptian, Babylonian and Greek Mathematics* (1954; trans. Arnold Dresden, 1961; rpt. New York: John Wiley, 1963), pp. 228 and 277.

7. For an alternative proof (likely Heron's own) based entirely on proportions, see Heilbron, *Geometry Civilized*, pp. 269–271.

8. Carl B. Boyer, *History of Analytic Geometry: Its Development from the Pyramids to the Heroic Age* (1956; rpt. Princeton Junction, N.J.: Scholar's Bookshelf, 1988), pp. 168–170. To quote Boyer, "This is possibly the first time that either of these formulae [the one given on page 134 and its three-dimensional counterpart] appeared in print. Consequently, credit for them should go, pending further evidence, to Clairaut." He adds, however, that Clairaut's contribution should not be exaggerated: "The distance formulae are, after all, obvious analytical expressions of an ancient theorem named for Pythagoras but known to the Babylonians of some four thousand years ago. There can be little doubt but that their equivalents were known to the earliest analytic geometers, including Descartes and Fermat."

9. I prefer the phrase "rectangular box" over the somewhat obsolete and tongue-breaking *parallelepiped*. (In one respected book I found it spelled *paralellelipiped*!)

10. Source: *Mathematical Adventures for Students and Amateurs*, ed. David F. Hayes and Tatiana Shubin (Washington, D.C.: Mathematical Association of America, 2004), p. 62.

11. The French mathematician D'Amondans Charles de Tinseau (1748–1822) in 1774 extended this result to *any* three-dimensional shape: the square of the area of any plane surface is equal to the sum of the squares of the projections of this surface on three mutually perpendicular coordinate planes. The special case of the tetrahedron was already known to Descartes. See Boyer, *History of Analytic Geometry*, p. 207.

12. For simplicity we ignore here the phase of each overtone relative to the fundamental; these phases do not as a rule affect the timbre of the sound.

A Pythagorean Curiosity

Toward the end of *The Pythagorean Proposition*, Loomis offered what he called "The Pythagorean Curiosity."[1] Starting with right triangle *ABC* (fig. S7.1), build the squares *BMNC*, *CDEA*, and *AHIB* on the legs *a*, *b* and the hypotenuse *c*, respectively. Draw *EH*, *IM*, and *ND*, and build the squares *EFGH*, *IKLM*, and *NPQD* on these lines. Finally, draw *LP*, *QF*, and *GK* and extend them until they meet at *A′*, *B′*, and *C′*. We have the following relations, which we state without proof. To shorten our notation, an equality sign will indicate that two line segments, or triangles or squares, have the same measure (length or area); the signs ∥ and ⊥ indicate that two lines are parallel or perpendicular, respectively, and the sign ~ means similarity.

1. Square *AHIB* = square *BMNC* + square *CDEA* (Euclid I 47).
2. △*AEH* = △*BIM* = △*CND* = △*CAB*.
3. *LP* ∥ *BC*, *QF* ∥ *CA*, and *GK* ∥ *AB*. Therefore, △*A′B′C′* ~ △*ABC*.
4. *LP* = 4*BC* = 4*a*, *QF* = 4*CA* = 4*b*, and *GK* = 4*AB* = 4*c*.
5. Trapezoid *LPNM* = trapezoid *QFED* = trapezoid *GKIH* = 5△*ABC*.
6. Square *EFGH* + square *IKLM* = 5 square *NPQD* = 5 square *AHIB*.
7. Line *C′C* bisects both right angles *C′* and *C*. Therefore, *C′C* ⊥ *ME*. (Note that *ME* is a line of symmetry for the entire construction.)
8. The square built on *GK* = square on *LP* + square on *QF* (these squares are not shown in the figure).

Loomis called these eight relations "demonstrable truths," and he added a ninth, which simply says, "Etc., etc.," hinting that more relations can be found. For example, the sides of △*A′B′C′* are to those of △*ABC* as the ratio $\frac{2(a+b)^2}{ab}$. And because △*ABC* and △*A′B′C′* are similar and oriented the same way, the entire construction can be repeated with △*A′B′C′* replacing △*ABC*, and so on ad infinitum, creating ever larger triangles with their attendant squares and trapezoids. Some curiosity!

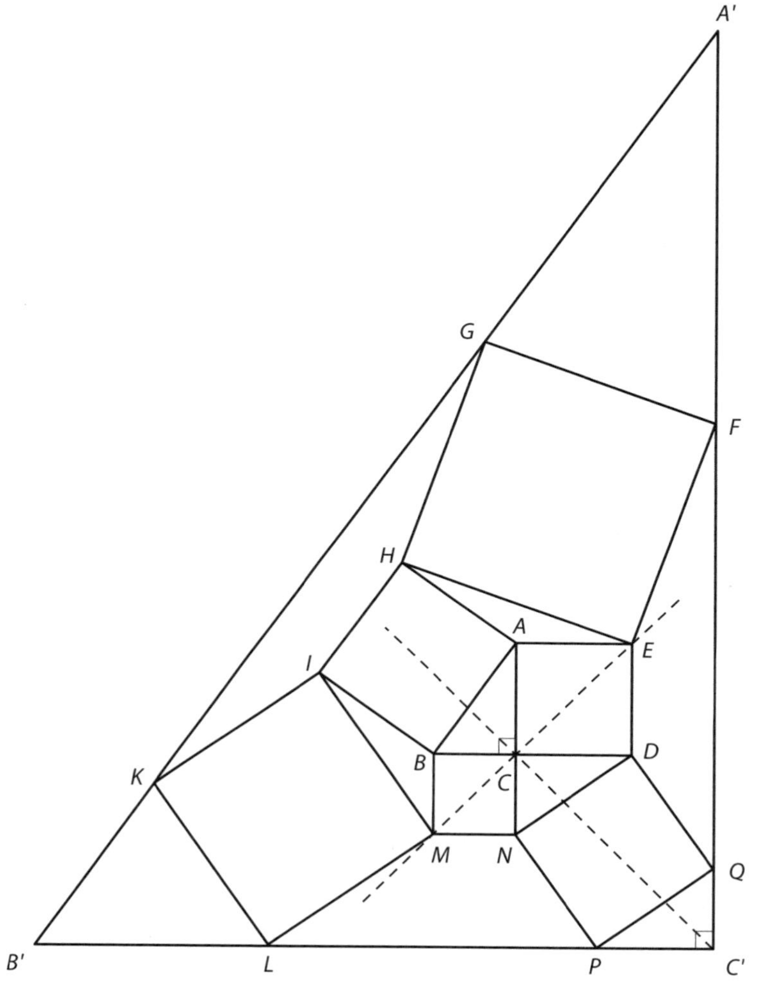

Figure S7.1. A "Pythagorean Curiosity"

Notes and Sources

1. Loomis, pp. 252–253. He traces the curiosity to a notebook of John Water-house, an engineer of New York City; it appeared in a New York paper in July 1899.

A Case of Overuse

It's all a matter of perspective.
—A common saying

 I don't usually pay much attention to newspaper ads, but this one in the *New York Times* of October 18, 2003, caught my eye (fig. S8.1). No, not the story about Bob, Mary, and Jack, but the rectangle above it, whose area the reader is asked to find—and explain how it was found.

Before you attempt to solve it, I should point out that a crucial piece of information was not given: the angle between the two slanted lines. It certainly *looks* as if the two lines form a right angle, but this was not stated; and without this information, the problem cannot be solved. So I had no choice but to assume that the two lines indeed form a 90° angle. I then followed the obvious path: the area of a rectangle is equal to its base times its height, so I first found the length of the base, using the Pythagorean theorem; that length is 5. Next I needed the height, so I called it h. Denoting the line segments DE and EC by x and y, respectively, we have $h^2 + x^2 = 3^2 = 9$ and $h^2 + y^2 = 4^2 = 16$. Subtracting the first equation from the second, we get $y^2 - x^2 = 7$. But $y^2 - x^2 = (y + x)(y - x)$, and $y + x$ is the length of the base, which we already found to be 5; so $5(y - x) = 7$, or $y - x = 7/5$. Solving the system of equations $y + x = 5$ and $y - x = 7/5$, we get $x = 9/5$, $y = 16/5$, from either of which find $h = 12/5$. The required area is thus $5 \cdot (12/5) = 12$.

Now by any standard, this would not exactly qualify as an elegant solution; it is a brute-force solution. But when I accidentally tilted the figure a bit, a much simpler solution suddenly sprang out of the page. From E drop a perpendicular to AB, meeting it at F (see fig. S8.2). Triangles ADE and AFE are congruent and thus have the same area; ditto for triangles BCE and BFE. Combined, these four triangles make up the rectangle. But triangles AFE and BFE add up to triangle AEB, whose area is simply $(3 \times 4)/2 = 6$ (thinking of AE as the base and BE as the height). So the area of the rectangle is twice as much, or 12. Hooray!

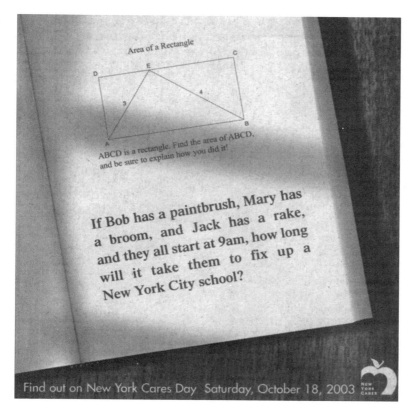

Figure S8.1 An ad in the *New York Times*

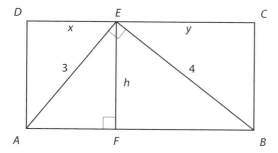

Figure S8.2. A simpler solution

This reminded me of a story I once heard from a friend. Years back, as a student, he was trying to prove a certain theorem in geometry. For hours he stared at the figure on the page in front of him, trying to discover some pattern in the complicated array of points and lines, but all in vain. Then a light breeze drifted through the open window and blew the page to the floor, where it landed upside down. And suddenly the proof popped up in front of his eyes! It took a change of perspective to make an intractable problem look almost trivial.[1]

Notes and Sources

1. Unfortunately my friend could not remember which theorem it was he was trying so hard to prove years back.

Strange Coordinates

A patu [beam?] of length 0;30 [stands against a
wall]. The upper end has slipped down a distance
0;6. How far did the lower end move?
    —BM 85 196 (a Babylonian text dating ca. 1800 BCE)

If you were shown the equation $\alpha^2 + \beta^2 = 1$, you would instantly recognize
it as the rectangular equation of the unit circle. The unit circle it is indeed; but
not in rectangular coordinates.

At the beginning of the nineteenth century, mainstream mathematics was
preoccupied with two broad subjects. In continental Europe a vast expansion
of the calculus was taking place, extending it to new areas such as ordinary
and partial differential equations, the theory of functions of a complex variable, the analysis of periodic functions, and mathematical physics. In England,
a century of stagnation following the death of Newton in 1727 came to an end
about 1830 with the birth of abstract algebra, a subject that would soon trans-
form the very nature of mathematics. Compared to these main trends, geometry was relegated to the sideline. There was a feeling that not much could be
added to what the Greeks had already accomplished two thousand years earlier. In addition, Descartes' invention of analytic, or coordinate geometry in
1637 had drastically changed the nature of geometry, in effect uniting it with
algebra. No more did one need a straightedge and compass to solve geometric
problems; these classic tools were now replaced by algebraic equations. And if
any proof was needed of the great power of analytic geometry, it was amply
supplied by the calculus, whose very foundation depended on the algebraic
description of lines, curves, and surfaces—so much so that nowadays the two
subjects are taught as one (at least in American colleges).

Not that geometry was totally stagnant. A small group of geometers,
mainly in France and Germany, revived the old interest in synthetic or "pure"
geometry, the kind of geometry that followed the Euclidean deductive approach. The basic tools of Euclidean construction were the straightedge (an
unmarked ruler) and compass; hundreds of constructions, many of them quite

complex, had been devised over the years using just these tools, elevating geometric construction to an art in its own right.

Yet when it came to regular polygons, the power of the straightedge and compass seemed to be limited: the only polygons that were known to be constructible with these tools were the regular 3, 4, 5, and 15 gons, and the polygons derived from them by doubling the number of sides. It came therefore as a total surprise when in 1796 the eighteen-year-old Carl Friedrich Gauss (1777–1855) proved that a regular polygon of seventeen sides can be constructed using the Euclidean tools. So impressed was young Gauss by this discovery that he decided to devote himself to mathematics, giving up his erstwhile love, linguistics; he soon became the world's leading mathematician of the first half of the nineteenth century, and is considered by many to rank with Archimedes and Newton among the three greatest mathematicians of all time. A statue of Gauss, standing on a seventeen-sided pedestal, honors him in his native town of Brunswick in Germany.[1]

A second surprise came just one year later, when an Italian geometer and poet, Lorenzo Mascheroni (1750–1800), in 1797 proved that every construction that can be done with a straightedge and compass can be achieved with a compass alone: the straightedge is not needed at all. (Of course, we cannot draw a straight line with a compass, but we can use a compass to determine the two points of intersection of two circles; and since two points determine a line uniquely, they are to be regarded as representing the line.)[2]

Gauss's and Mascheroni's discoveries showed that good old classical geometry was far from exhausted. Indeed, one branch of geometry that was markedly different from Euclid's had already been known for over a century: projective geometry. This beautiful but esoteric subject had its genesis in the sixteenth century, when there was a great deal of interest in the new art of perspective. When an artist depicts a scene on canvas, some features, such as the shape of objects or their relative size, appear distorted, whereas others remain unchanged. As an example of the latter, consider three points on a line: their images on the canvas will still lie on a line, assuming, of course, that the artist is following the laws of perspective. Projective geometry is the study of those properties of a figure that remain unchanged—*invariant*, in the language of mathematics—when projected onto the canvas. This shift in emphasis from the *metric* properties of a figure (the length of a line segment, the angle between two lines, or the area of a polygon) to properties of *incidence* (the position of points, lines, and planes relative to one another) marked the first significant departure from Euclidean geometry in over two thousand years.[3]

Central to projective geometry is the concept of *duality*: the fact that, as far as incidence is concerned, there is a complete equivalence between points and lines in the plane (as also between points and planes in space). Take, for example, the statement, *two points determine one and only one line*. If we interchange the words "point" and "line," the statement reads *two lines determine*

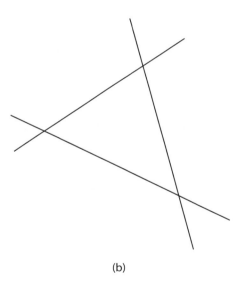

(a)

(b)

Figure 10.1. Dual views of a triangle: (a) as three noncollinear points; (b) as three nonconcurrent lines

one and only one point.[4] According to the principle of duality, any valid statement about the mutual positions of points and lines remains valid if the words "point" and "line" are interchanged. For example, a triangle can be thought of as a set of three noncollinear points (points not lying on one line), or as a set of three nonconcurrent lines (lines not intersecting at one point), as shown in figure 10.1. The former is the more common definition, but the latter is just as valid (of course, we must think of the lines as extending indefinitely, which gives the triangle a somewhat strange look).

The principle of duality is one of the most elegant concepts in mathematics, because of its power to unite statements that at first glance seem quite unrelated. Indeed, in older geometry books one can find pages neatly divided by vertical lines, with a statement and its dual appearing side by side across the line. But this elegance comes at a price: such common notions as the distance between two points, the angle between two intersecting lines, or the area of a closed figure—in short, those properties that can be assigned numerical values—have no place in projective geometry. And that includes the Pythagorean theorem.

Or so at least it seemed until 1828, when a German mathematician by the name of Julius Plücker (1801–1868) brought it back through the rear door. Plücker took the principle of duality to the extreme. If, he argued, points and lines are completely equivalent, why can't lines be used to construct curves in much the same way as points are? Just as we can think of a curve as the locus of points with a common property, so can we think of it as the locus of lines tangent to it (fig. 10.2). In this interpretation, the curve is the *envelope of its tangent lines*. For example, we usually think of a circle as the set of all points equidistant from the center, but we can just as well think of it as the set of all tangent lines equidistant from the center (fig. 10.3). It is true that the first interpretation is easier to construct geometrically (you only need a compass), but the second interpretation is just as valid.

But Plücker went a step further: he gave this idea an analytic formulation by introducing a new type of coordinates he called *line coordinates*. Start with the equation of a line in the *xy* plane, $Ax + By = C$, where A and B are not both 0. As long as $C \neq 0$ (that is, as long as the line does not pass through the origin), we can divide the equation by C to get

$$\alpha x + \beta y = 1, \tag{1}$$

where $\alpha = A/C$, $\beta = B/C$. Plücker was struck by the complete symmetry of this equation in regard to x and y on the one hand, and α and β on the other. We usually think of α and β as constants, whereas x and y are variables. Equation (1) then describes all points $P(x, y)$ lying on the line determined by the constants (α, β). Because these constants determine the line uniquely, we may regard them as the (fixed) coordinates of the line and designate it as $\ell(\alpha, \beta)$. But in view of the complete symmetry between the variables (x, y) and the constants (α, β) in equation (1), we may interchange their roles and regard (x, y) as fixed and (α, β) as variables. Equation (1) then describes *all lines $\ell(\alpha, \beta)$ passing through the fixed point $P(x, y)$*. To repeat: equation (1) can be equally

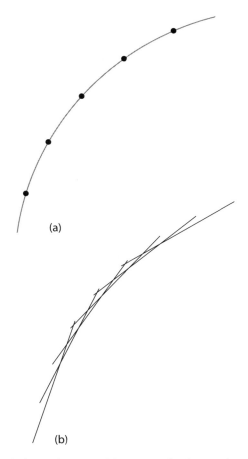

(a)

(b)

Figure 10.2. Dual views of a curve: (a) as a set of points on the curve; (b) as a set of lines tangent to the curve

interpreted as the equation of a line ℓ whose fixed coordinates are (α, β), or as the equation of a point P whose fixed coordinates are (x, y). These dual interpretations are shown in figure 10.4.

However, In order to achieve complete duality between points and lines, we must ask whether a geometric meaning can be given to the line coordinates (α, β) similar to the meaning of the point coordinates (x, y) as the distances of P from the y- and x-axes. Such an interpretation does indeed exist. If we set $y = 0$ in equation (1), we get $x = 1/\alpha$; similarly, by setting $x = 0$, we get $y = 1/\beta$. But the values of x and y so found are the intercepts m and n of the line with the x- and y-axes, respectively. Thus,

$$\alpha = 1/m, \quad \beta = 1/n, \tag{2}$$

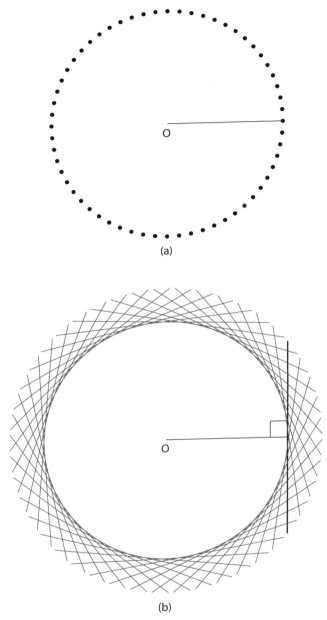

(a)

(b)

Figure 10.3. Dual views of a circle (a) as a set of points equidistant from 0; (b) as a set of lines equidistant from 0

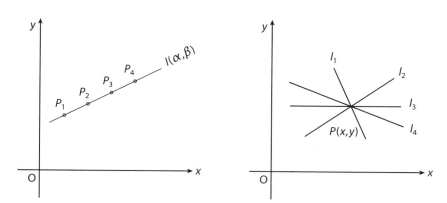

(a)   (b)

Figure 10.4. Dual interpretation of $\alpha x + \beta y = 1$: (a) as the equation of a line; (b) as the equation of a point.

and the line coordinates are seen to be the reciprocals of the corresponding intercepts with the axes (fig. 10.5).

At first glance line coordinates might seem a strange construct, but we should realize that the goal of any coordinate system is to determine uniquely the position of the objects for which it is designed—whether points, lines, or any other geometric entity—and do so in the simplest possible way. When we encounter polar coordinates for the first time, they too look strange, but in many situations they are preferable to the more common rectangular (Cartesian) coordinates: just think of the way an air traffic controller locates an aircraft on the radar screen in terms of the plane's distance and direction from the tower. Moreover, quite often the equation of a curve becomes simpler when expressed in polar instead of rectangular coordinates. For example, the rectangular equation of the unit circle is $x^2 + y^2 = 1$, whereas its polar equation is the much simpler $r = 1$.

To stress this point, let us examine a case that actually calls for the use of line coordinates. A ladder of length 1 is leaning against a wall, with its bottom end free to slide along the floor in a direction perpendicular to the wall. As the ladder assumes all possible positions, what is the shape of the region it occupies? Referring to figure 10.6, let m be the distance of the bottom end of the ladder from the wall, and n the height of the top end from the floor. As the bottom end slides away from the wall, the ladder slowly turns in the xy plane, tracing a curve to which it is always tangent. Our goal is to find the equation of this curve.

By the Pythagorean theorem, the equation $m^2 + n^2 = 1$ must always be fulfilled, regardless of the position of the ladder. But m and n are the intercepts of

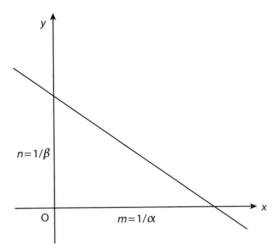

Figure 10.5. Line coordinates

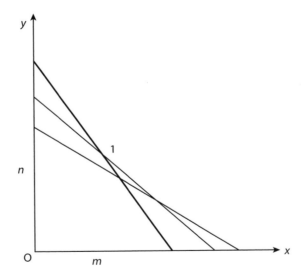

Figure 10.6. A sliding ladder

the (variable) tangent line to the curve, so we have m = 1/α, n = 1/β, where α and β are the line coordinates of the tangent line. Putting this back into $m^2 + n^2 = 1$, we get

$$\frac{1}{\alpha^2} + \frac{1}{\beta^2} = 1. \tag{3}$$

As the ladder assumes all possible positions, each tangent line is determined by its line coordinates (α, β) in accordance with equation (3). Equation (3) is therefore the line equation of the required curve; it completely determines the curve and is thus the solution to the problem we have set out to solve.

Still, most of us, conditioned as we are to thinking in terms of rectangular or "point" coordinates, will feel uncomfortable with this solution. Can we translate equation (3) into a rectangular, point equation? (To avoid possible misunderstanding, we stress that $m^2 + n^2 = 1$ is *not* the required equation, because m and n are not the coordinates of a point on the curve.) The answer is yes, but the process is a bit tedious and requires some calculus, so I will relegate it to Appendix G. The resulting equation turns out to be

$$x^{2/3} + y^{2/3} = 1, \qquad (4)$$

whose graph is the star-shaped *astroid* we met in chapter 7.[5]

As a sequel to this example, here is a problem I like to call "the mover's dilemma." A moving company needs to carry a long object, say a sofa of length a, through an L-shaped hallway (fig. 10.7). Will the sofa get through? Let us choose the x- and y-axes along the outer walls, and let the two branches of the hallway have widths p and q; the corner of the inner wall then has coordinates (p, q). If the sofa is relatively short, there is no problem carrying it around the corner. But if it is long enough, it might get stuck. As the sofa is maneuvered around the corner, it traces the astroid given by the equation $x^{2/3} + y^{2/3} = a^{2/3}$.[6] Any point (x, y) lying *inside* the region described by this astroid will be struck by the sofa, while points outside this region will be safe. Thus, for the sofa to clear the corner, we must have $p^{2/3} + q^{2/3} \geq a^{2/3}$. In the special case when $p = q$ (the two branches of the hallway having the same width), this condition reduces to $2p^{2/3} \geq a^{2/3}$, from which we get $p \geq a/\sqrt{8}$ ~ $0.35\, a$; if the hall is any narrower, the sofa will get stuck. This also shows that the two corners must be separated by a distance of at least $p\sqrt{2} = a/2$.[7]

At the beginning of this chapter I mentioned the equation $\alpha^2 + \beta^2 = 1$. Given that α and β represent line coordinates, we naturally wish to find out what curve this equation represents. The answer is, *the unit circle, generated by the set of its tangent lines.* To see this, we recall from chapter 9 that in any right triangle, the length d of the perpendicular from the right angle to the hypotenuse satisfies the equation $1/d^2 = 1/a^2 + 1/b^2$. Referring to figure 10.8 and denoting the perpendicular distance from the origin to the tangent line by p, we have

$$1/p^2 = 1/m^2 + 1/n^2, \qquad (5)$$

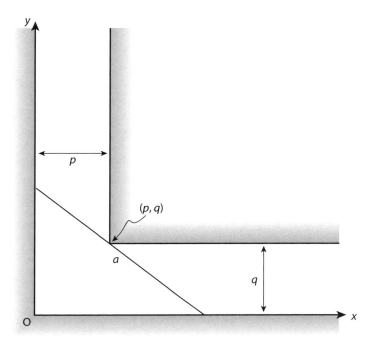

Figure 10.7. The mover's challenge: will the sofa clear the corner?

where *m* and *n* are the *x*- and *y*-intercepts of the line, respectively. Replacing *m* and *n* by $1/\alpha^2$ and $1/\beta^2$ and putting $p = 1$ for the unit circle, we get $\alpha^2 + \beta^2 = 1$, the line equation of the unit circle.

❖   ❖   ❖

Plücker's line coordinates are a fine example of how duality plays a role in mathematics. Regrettably, they are all but forgotten today, but they did make a comeback of sorts in the art of line designs—geometric patterns generated entirely by straight lines strung according to set rules (see fig. 10.9).[8] As for Plücker, his career took a strange twist. After publishing his ideas in his major work, *Developments in Analytic Geometry* (in two volumes, 1828 and 1831), he abruptly abandoned mathematics and turned to experimental physics. For eighteen years, from 1846 to 1864, he investigated the magnetic properties of crystals, helped in building a new standard thermometer, and studied the spectral lines of gases, a subject that would soon become the backbone of astrophysics. Then, just as abruptly, he returned to his erstwhile love, spending his last four years further developing line coordinates. He died in 1868 at the age of sixty-seven. It is a pity that his name and work have almost entirely disappeared from today's geometry curriculum.

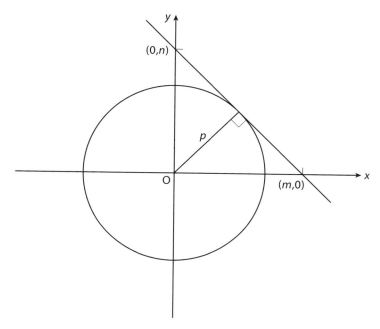

Figure 10.8. Deriving the equation $\alpha^2 + \beta^2 = 1$

Notes and Sources

Note: The epigraph is quoted in Bartel L. van der Waerden, *Science Awakening: Egyptian, Babylonian and Greek Mathematics* (1954; trans. Arnold Dresden, 1961; rpt. New York: John Wiley, 1963), p. 76. The numbers are in the sexagesimal (base 60) numeration system. The document is another proof that the Babylonians already knew the Pythagorean theorem at least a thousand years before Pythagoras.

1. Gauss actually went farther. He showed that a regular polygon with a prime number of sides can be constructed with a straightedge and compass if this prime number is of the form $N = 2^{2^n} + 1$, where n is a nonnegative integer. For $n = 0$, 1, 2, 3, and 4 we get $N = 3$, 5, 17, 257, and 65,537—all primes.

Primes of the form $2^{2^n} + 1$ are called *Fermat primes*; Pierre de Fermat in 1654 conjectured that $2^{2^n} + 1$ is a prime for every nonnegative integer n, but this was disproved in 1732, when Leonhard Euler showed that for $n = 5$ we get the composite number $4{,}294{,}967{,}297 = 641 \times 6{,}700{,}417$. It is not known whether any additional Fermat primes exist; thus it may be that there are as yet undiscovered regular polygons constructible with the Euclidean tools. However, if such polygons exist, they have a huge number of sides, making any *practical* construction totally out of the question.

The proof that Fermat primes are the *only* primes for which the construction is possible was given in 1837 by Pierre Laurent Wantzel (1814–1848); thus Gauss's condition is both necessary and sufficient.

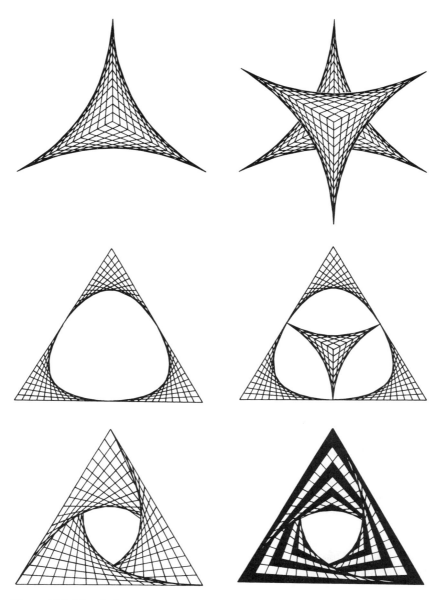

Figure 10.9. Line designs

2. In 1928 a student of the Danish mathematician J. Hjelmslev (1873–1950) discovered in a Copenhagen bookstore a volume entitled *Euclides Danicus*, published in 1672 by an obscure German geometer, Georg Mohr (1640–1697). Upon examining the book, he was surprised to find in it a complete proof of Mascheroni's result, written 125 years before Mascheroni. More about Mascheroni constructions can be found in Richard Courant and Herbert Robbins, *What Is Mathematics?* (1941; revised by Ian Stewart, New York: Oxford University Press, 1996), pp. 147–152.

3. For a good introduction to projective geometry, see ibid., chap. 4.

4. If the lines are parallel, they "intersect" at a point at infinity. The admittance of points and lines at infinity (also called *ideal points* and *lines*) as legitimate geometric objects is a central tenet of projective geometry; see *To Infinity and Beyond*, chap. 15.

5. The astroid has many other interesting properties. See *Trigonometric Delights*, pp. 98–99, 100–101, and 106. See also Robert C. Yates, *Curves and Their Properties* (Reston, Va.: National Council of Teachers of Mathematics, 1974), pp. 1–3. For more on line equations, see my article, "Line Equations of Curves: Duality in Analytic Geometry," *International Journal of Mathematics Education in Science and Technology*, vol. 2, no. 3 (1978), from which parts of this chapter are adapted.

6. This is a generalization of the equation $x^{2/3} + y^{2/3} = 1$ for the case where the generating rod has length a rather than 1; it is analogous to the equation $x^2 + y^2 = a^2$ of a circle with center at O and radius a, as compared to the equation of the unit circle, $x^2 + y^2 = 1$.

7. We have tacitly assumed that the problem is two-dimensional, so that the sofa cannot be tilted vertically. This problem would have fit nicely into Edwin A. Abbott's classic 1884 science fiction story *Flatland: A Romance of Many Dimensions* (rpt. Princeton, N.J.: Princeton University Press, 2005).

8. See, for example, Dale Seymour, Linda Silvey, and Joyce Snider, *Line Designs* (Palo Alto, Calif.: Creative Publications, 1974).

Notation, Notation, Notation

Whatever you say to mathematicians they
translate into their own language, and forthwith it
is something entirely different.
—Johann Wolfgang von Goethe (1749–1832)

Mathematics and music share many traits, not least of which being their re-
liance on a good system of notation. The ancients were entirely dependent on
verbal, "do such and such" instructions: multiply two given numbers, play two
specified notes, and the like. Needless to say, such verbal instructions were
vague and inefficient. In mathematics, in particular, the lack of a good system
of notation hampered the Greeks from making inroads into areas other than
arithmetic and geometry.

It took over a thousand years until mathematics made the transition from
verbal to *literal algebra*, in which letters and symbols replaced word-of-
mouth instructions. This transition began in the 1400s and reached maturity
around 1600, when François Viète introduced a system of notation in which
consonants stood for known quantities and vowels for unknowns (see p. 76).
The idea of using letters to denote algebraic quantities, so natural to us today,
was a novelty in his time, and it would greatly facilitate the formulation of
mathematical statements. But Viète, perhaps taken aback by the audacity of
his own innovation, did not push it all the way through. Although he used the
modern symbols + and − for addition and subtraction, he wrote *aequatur* for
equality, and *A quadratus* and *A cubus* for a^2 and a^3 (though he later abbrevi-
ated them to Aq and Ac). For the equation $a^2 + b^2 = c^2$ he wrote $Aq + Bq$ *aequatur Cq*—still not as concise as our modern notation, but coming close.

With Descartes, the transition to symbolic algebra was by and large complete. (Interestingly, Newton, living half a century after Descartes, still occasionally wrote *aa* for a^2, but for higher powers he used our modern exponential notation.) Soon, however, the accelerating pace of mathematical discoveries
demanded new symbols and new rules to handle them. In 1843 the Irish mathematician Rowan Hamilton (1805–1865), in an attempt to extend ordinary
complex numbers to three dimensions, invented *quaternions*, abstract objects

that could be added and subtracted according to the familiar rules of arithmetic, but which defied the commutative law of multiplication, $ab = ba$. A quaternion is an expression of the form $q = a + bi + cj + dk$, where the four *unit quaternions* 1, i, j, and k follow the multiplication laws $i \times j = -j \times i = k$, $j \times k = -k \times j = i$, $k \times i = -i \times k = j$, and $i^2 = j^2 = k^2 = i \times j \times k = -1$. Associated with each quaternion is its *magnitude* or *absolute value*, $|q| = \sqrt{a^2 + b^2 + c^2 + d^2}$. This expression can be interpreted as the "distance" of the quaternion from the origin, as follows from the Pythagorean theorem in four-dimensional space.

Hamilton is said to have discovered the multiplication laws of quaternions in a flash while crossing Brougham Bridge in his native Dublin, where a plaque commemorates the occasion (fig. 11.1). This event marked the beginning of abstract algebra, the realization that mathematics need not be limited to the description of "real" objects like numbers or geometric quantities. On the contrary, it could be extended to arbitrary constructs, subject to formal rules of operation, as long as these rules are free from internal contradictions.

As ingenious as quaternions were, the fact that they required four components to represent three-dimensional space made them somewhat awkward to use. They were soon superseded by a new concept, *vectors*, developed chiefly by the American physicist Josiah Willard Gibbs (1839–1903). As conceived originally, a vector represented a physical quantity having both magnitude and direction; familiar examples are force and velocity. Soon this concrete picture gave way to the idea of a *directed line segment*, denoted simply by PQ. This concise notation follows the simple rules $PQ = -QP$ and $PQ + QR = PR$, the latter expressing the fact that if you go from point P to point Q and then proceed to point R, the end result is the same as if you went directly from P to R. To make the system complete, we define the *zero vector* as the vector from P to itself; that is, $\mathbf{0} = PP$; we then have $PQ + QP = PP = \mathbf{0}$. These rules allow us to manipulate vectors as if they were purely algebraic objects, regardless of the physical quantity they might represent.

Several vector notations have evolved over the years and are being used today, sometimes with confusing results. One still finds the notation $\overrightarrow{PQ}$ for the line segment from P to Q, though the little arrow is redundant: the order of the letters in PQ makes it clear that the vector points from P to Q. Often a single lower-case bold letter is used to denote a vector, as in $\mathbf{a} = PQ$. To denote the magnitude of PQ we use absolute value bars, as in $|PQ|$, or if the vector is denoted by $\mathbf{a}$, the same letter is set in plain type. For the zero vector we write $\mathbf{0}$; clearly, $|\mathbf{0}| = 0$.

The rule $PQ + QR = PR$ is merely the familiar *triangle rule for adding vectors* (also known as the *parallelogram rule*; see fig. 11.2). Note, however, that the rule does *not* apply to the magnitudes of the respective vectors. The shortest distance between two points is the straight line segment connecting them; so if you wish to go from P to R but decide to make an intermediate stop at Q,

Figure 11.1. The plaque on Brougham Bridge, Dublin

this detour can only lengthen your trip (unless Q happens to be on the line segment PR). Translated into vector language, this means that $|PQ| + |QR| \geq |PR|$; and since PR is the vector sum of PQ and QR, we have the *triangle inequality*

$$|PQ| + |QR| \geq |PQ + QR|,$$

an important relation that is usually written right-to-left; using lower-case bold letters, it reads

$$|\mathbf{a} + \mathbf{b}| \leq |\mathbf{a}| + |\mathbf{b}|. \tag{1}$$

The equality sign holds if, and only if, $\mathbf{a}$ and $\mathbf{b}$ are on the same line and have the same direction.

When two vectors $\mathbf{a}$ and $\mathbf{b}$ are perpendicular to each other, their sum $\mathbf{a} + \mathbf{b}$ is the hypotenuse of a right triangle whose legs are $\mathbf{a}$ and $\mathbf{b}$ (fig. 11.3). The Pythagorean theorem then says that

$$|\mathbf{a} + \mathbf{b}|^2 = |\mathbf{a}|^2 + |\mathbf{b}|^2, \tag{2}$$

which should not be confused with the identity $(a + b)^2 = a^2 + 2ab + b^2$ familiar to us from elementary algebra.

If the initial point of a vector is at the origin O and its terminal point at P, it is called the *radius vector* of P and is denoted by $\mathbf{r}$; that is, $\mathbf{r} = OP$. In two dimensions, let the rectangular coordinates of P be (x, y); we can then write $\mathbf{r}$ as a linear combination of its components, $\mathbf{r} = x\mathbf{i} + y\mathbf{j}$, where $\mathbf{i}$ and $\mathbf{j}$ are the *unit*

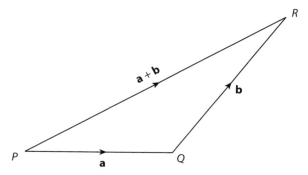

Figure 11.2. Vector addition

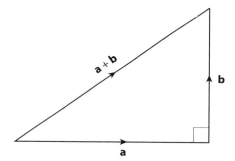

Figure 11.3. The sum of two perpendicular vectors

vectors along the *x*- and *y*-axes, respectively (fig. 11.4). This short notation can be shortened still more by writing $\mathbf{r} = (x, y)$, in effect identifying the radius vector of *P* with *P* itself. The magnitude of $\mathbf{r}$ is then simply the distance of *P* from the origin; that is, $r = \sqrt{x^2 + y^2}$. This formula, of course, can be extended to three dimensions, in which case it becomes $r = \sqrt{x^2 + y^2 + z^2}$. In what follows we will refer to this distance as the *length* of $\mathbf{r}$.

The introduction of vectors into mathematics greatly facilitated the description of various physical laws. For example, Newton's second law of motion, $F = ma$, is really a relation between the force vector $\mathbf{F}$ acting on a mass *m*, and the acceleration $\mathbf{a}$ imparted to it by this force: $\mathbf{F} = m\mathbf{a}$. Had we not used vector notation, we would have to write three separate equations, one for each component of the force and acceleration vectors ($F_x = ma_x$, and similarly for *y* and *z*).

The great utility of this notation has made vector algebra—and later, vector calculus—an indispensable tool in physics. But mathematicians went one step

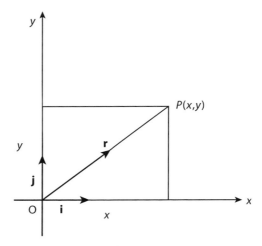

Figure 11.4. A vector and its rectangular components

farther: they turned the table and defined *any* ordered set of *n* numbers a *vector in n-dimensional space*. The actual nature of these numbers does not really matter—they could be variables, coordinates, components of some physical quantity, or just pure numbers; the array $(x_1, x_2, \ldots, x_n)$ itself is what mattered. Mathematicians routinely speak of the "length" of such a vector, even if they do not think of it in terms of inches or feet. This length is given by the formula $\sqrt{x_1^2 + x_2^2 + \cdots + x_n^2}$; it can be extended to an infinitely many dimensional space $(x_1, x_2, \ldots)$, provided the sum $x_1^2 + x_2^2 + \ldots$ converges.

As always in mathematics, when introducing a new object, we must first set the rules that govern this object—the rules of the game, so to speak. In the case of vectors, these rules are the "natural" ones: two vectors are equal if and only if their corresponding components are equal; that is, $\mathbf{a} = \mathbf{b}$ if and only if $a_i = b_i$ for all $i = 1, 2, \ldots, n$. The sum of two vectors $\mathbf{a} = (a_1, a_2, \ldots, a_n)$ and $\mathbf{b} = (b_1, b_2, \ldots, b_n)$ is the vector $\mathbf{a} + \mathbf{b} = (a_1 + b_1, a_2 + b_2, \ldots, a_n + b_n)$. And the product of a number c (a "scalar") and a vector $\mathbf{a}$ is the vector $c\mathbf{a} = (cx_1, cx_2, \ldots, cx_n)$. Geometrically, this last operation "stretches" the length of $\mathbf{a}$ in the ratio $c{:}1$ (if $c < 0$, it also reverses the direction of $\mathbf{a}$). As an example, let $\mathbf{a} = (2, -3)$ and $b = (3, 4)$. Then $\mathbf{a} + \mathbf{b} = (5, 1)$ and $7\mathbf{a} = (14, -21)$. These operations follow the familiar laws of arithmetic, specifically the commutative, associative, and distributive laws, as shown in figure 11.5.

But what about *multiplying* two vectors? This, too, can be defined, but the result is not a vector but a scalar, a number. Hence this product is sometimes called a *scalar product*, although mathematicians prefer the names *inner product* or

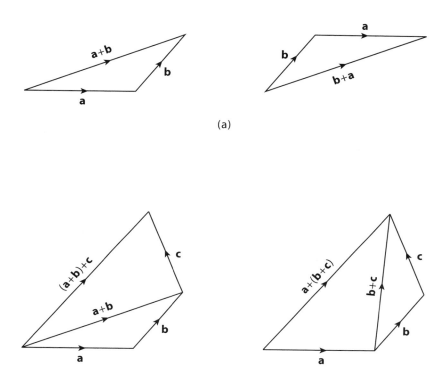

(a)

(b)

Figure 11.5. Vector algebra: (a) $\mathbf{a} + \mathbf{b} = \mathbf{b} + \mathbf{a}$; (b) $(\mathbf{a} + \mathbf{b}) + \mathbf{c} = \mathbf{a} + (\mathbf{b} + \mathbf{c})$

dot product. The dot product of the vectors $\mathbf{a} = (a_1, a_2, \ldots, a_n)$ and $\mathbf{b} = (b_1, b_2, \ldots, b_n)$ is the *number* $a_1 b_1 + a_2 b_2 + \ldots + a_n b_n$. This product is denoted by $\mathbf{a} \cdot \mathbf{b}$ (hence the "dot in the name). For example, if $\mathbf{a} = (2, -3)$ and $\mathbf{b} = (3, 4)$, then $\mathbf{a} \cdot \mathbf{b} = 2 \times 3 + (-3) \times 4 = -6$. Using the Greek letter Σ (sigma) to denote summation, we can write the dot product as:

$$\mathbf{a} \cdot \mathbf{b} = \sum_{i=1}^{n} a_i b_i. \tag{3}$$

When we take the dot product of a vector $\mathbf{a} = (a_1, a_2, \ldots, a_n)$ with itself, we get $\mathbf{a} \cdot \mathbf{a} = a_1^2 + a_2^2 + \cdots + a_n^2 = \sum_{i=1}^{n} a_i^2$. But this sum is precisely the square of the length of $\mathbf{a}$. We therefore have yet another characterization for the length of a vector:

$$|\mathbf{a}| = \sqrt{\mathbf{a} \cdot \mathbf{a}}. \tag{4}$$

If we square both sides of equation (4) and use it with $\mathbf{a} + \mathbf{b}$ replacing $\mathbf{a}$, we get

$$|\mathbf{a} + \mathbf{b}|^2 = (\mathbf{a} + \mathbf{b}) \cdot (\mathbf{a} + \mathbf{b}) = \mathbf{a} \cdot \mathbf{a} + \mathbf{a} \cdot \mathbf{b} + \mathbf{b} \cdot \mathbf{a} + \mathbf{b} \cdot \mathbf{b}$$
$$= \mathbf{a} \cdot \mathbf{a} + 2\,(\mathbf{a} \cdot \mathbf{b}) + \mathbf{b} \cdot \mathbf{b} = |\mathbf{a}|^2 + 2\,(\mathbf{a} \cdot \mathbf{b}) + |\mathbf{b}|^2, \tag{5}$$

where we have used the commutative and distributive laws to open the parentheses and collect like terms. Assume now that $\mathbf{a}$ and $\mathbf{b}$ are perpendicular to one another. Recalling equation (2) and comparing it with equation (5), we conclude that in this case $\mathbf{a} \cdot \mathbf{b} = 0$. This conclusion also works in reverse: if $\mathbf{a} \cdot \mathbf{b} = 0$, the vectors are perpendicular or *orthogonal*, in the language of vector algebra. Thus,

$$\mathbf{a} \perp \mathbf{b} \Leftrightarrow \mathbf{a} \cdot \mathbf{b} = 0 \tag{6}$$

(the symbol $\Leftrightarrow$ is read "if and only if"). And because $\mathbf{a} \cdot \mathbf{b} = \sum_{i=1}^{n} a_i b_i$ we can write this statement in the alternative form,

$$\mathbf{a} \perp \mathbf{b} \Leftrightarrow \sum_{i=1}^{n} a_i b_i = 0. \tag{7}$$

Note that this is true regardless of the number of dimensions of our space. As an example, let $\mathbf{a} = (1,2,-3)$ and $\mathbf{b} = (4,7,6)$. Then $\mathbf{a} \cdot \mathbf{b} = 1 \times 4 + 2 \times 7 + (-3) \times 6 = 0$, showing that $\mathbf{a}$ and $\mathbf{b}$ are orthogonal. To arrive at this conclusion more traditionally, we would have to do some pretty tedious calculations with the distance formula. This demonstrates the advantage of vector algebra over traditional (nonvector) methods.

A curious aside here is that Albert Einstein, in his early work on the general theory of relativity, managed to shorten the notation $\mathbf{a} \cdot \mathbf{b} = \sum_{i=1}^{n} a_i b_i$ even further. He realized that the sigma symbol is really redundant: it is implied by the fact that the summation index i in $\sum_{i=1}^{n} a_i b_i$ appears twice. According to Einstein's convention, whenever a summation index appears twice in a term, that term is to be summed over the relevant range (with the sigma symbol dropped). In this notation, the Pythagorean theorem takes the form $|\mathbf{a}|^2 = a_i a_i = a_i^2$. Einstein's summation convention is often used in advanced textbooks on mathematical physics, but algebra and calculus texts usually stick to the sigma notation.

One would think that by now the different ways of saying $a^2 + b^2 = c^2$ have been exhausted, but not so: the mathematician's appetite for generalizing an erstwhile "concrete" concept apparently knows no limit. Our vectors so far were n-dimensional objects, n-*tuples* of the form $(x_1, x_2, \ldots, x_n)$. Even if we

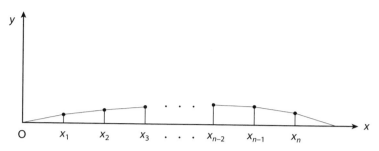

Figure 11.6. An approximation to a vibrating string

let $n -> \infty$, the number of dimensions is still countable, in the sense that we can match it with the natural numbers 1, 2, 3, ... But consider now a vibrating violin string. As a crude approximation, we can think of the string as consisting of n small point-masses distributed at points $x_1, x_2, \ldots, x_n$ along the length of the string (fig. 11.6). Each mass vibrates up and down in a direction parallel to the y-axis. Let us denote the deviation of the i-th mass from its equilibrium position by y_i; this gives us a set of n coordinates $(y_1, y_2, \ldots, y_n)$, a vector in n dimensions. The motion of the system as a whole is determined once we know the values of these coordinates at every instant.

But this was only an approximation, because a real string has its mass distributed *continuously* along its length. So instead of n discrete variables $(y_1, y_2, \ldots, y_n)$ we now have a *function* $y = f(x)$, where x and y are continuous variables. We can still think of $f(x)$ as a vector, but it is a vector whose number of dimensions is not only infinite, but *uncountable*, the kind of infinity of the number continuum. We can ascribe to this vector all the properties of an n-dimensional vector; in particular, we can assign it a *length*. This, however, requires us to make the transition from a discrete to a continuous sum, an *integral*. Just as in the discrete case we defined the length of a vector **a** as $\sqrt{\Sigma_{i=1}^n a_i^2}$ (see p. 163), so can we define the "length" of a function $f(x)$ as $\sqrt{\int_a^b [f(x)]^2 \, dx}$ (where a and b are the endpoints of the interval under consideration, in this case the length of the string)—provided the integral under the square root exists. The expression $\sqrt{\int_a^b [f(x)]^2 \, dx}$ is called the *norm* of $f(x)$ and is denoted by $\|f(x)\|$ to distinguish it from the ordinary absolute value sign $|f(x)|$. To recap:

$$\text{norm of } f(x) = \|f(x)\| = \sqrt{\int_a^b [f(x)]^2 \, dx}. \tag{8}$$

Not only can we ascribe a norm to a single function, but we can do so to the members of an entire set of functions, a *function space*, in which functions $f(x)$, $g(x), \ldots$ take the place of vectors **a**, **b**, $\ldots$. This space, defined on an interval $[a, b]$ and subject to the condition that $\int_a^b [f(x)]^2 \, dx$ should be finite, is called a

Hilbert space, after the eminent German mathematician David Hilbert (1862–1943).[1] A Hilbert space obeys all the formal rules of an ordinary vector space. Corresponding to the dot product $\mathbf{a} \cdot \mathbf{b}$ we now have the *inner product* (f, g) of two functions $f(x)$ and $g(x)$, defined as $\int_a^b f(x)g(x)dx$ (note the similarity of this definition to equation (3), with the integral replacing the sigma sign). Moreover, $f(x)$ and $g(x)$ must obey the triangle inequality

$$\|f(x) + g(x)\| \leq \|f(x)\| + \|g(x)\|. \tag{9}$$

Two functions are said to be *orthogonal* if $(f, g) = 0$, just as two ordinary vectors are perpendicular to each other if $\mathbf{a} \cdot \mathbf{b} = 0$. And the Pythagorean theorem? Why, it is expressed by the equation

$$\|f(x) + g(x)\|^2 = \|f(x)\|^2 + \|g(x)\|^2, \tag{10}$$

in complete analogy with equation (2) for an n-dimensional vector space. We can even define the "distance" between two functions as $\sqrt{\int_a^b [f(x) - g(x)]^2 dx}$. The keen reader will recognize in this expression the ghost of the familiar distance formula $\sqrt{(x_2 - x_1)^2 + (y_2 - y_1)^2}$.

At first thought, such a sweeping generalization of the vector concept may seem far removed from reality. What do we mean by saying that two functions are "perpendicular" (orthogonal) to each other? How can we envision a right triangle whose "sides" are functions? Such questions arise whenever an erstwhile concrete concept is expanded to new realms; there comes to mind the controversy about the meaning of the "imaginary" quantity $\sqrt{-1}$ when it first appeared on the mathematical scene.

Yet there are situations that call for just this kind of generalization. Consider again a violin string. As we saw in chapter 9, a string can vibrate not only in one mode, but in infinitely many modes whose frequencies are integral multiples of the lowest, fundamental frequency (fig. 11.7). Each mode has the shape of a sine wave, so we can describe it by a function of the form $f_n(x) = a_n \sin nx$, $n = 1, 2, 3, \ldots$ (for simplicity, we take the string to extend from $x = 0$ to $x = \pi$). Each of these functions represents the amplitude of a point x as it vibrates up and down in a particular mode. Now the energy of vibrations is proportional to the square of the amplitude, so the total energy content of each mode is proportional to $\int_0^\pi [f_n(x)]^2 dx$. And since this energy can come only from the initial energy put into the string when it was plucked, the value of $\int_0^\pi [f_n(x)]^2 dx$ must be finite. But this integral is precisely the square of the norm of $f_n(x)$ (see equation (8)). It can also be shown that the triangle inequality (equation (9)) holds for every pair of functions $f_m(x)$ and $f_n(x)$, and that $(f_m, f_n) = 0$ whenever $m \neq n$; that is, any two different functions in the set are orthogonal. The set of functions $\{f_n(x)\}$ is therefore a Hilbert space.[2]

Hilbert spaces turned out to play a key role in several areas of modern mathematical physics, among them functional analysis, differential equations, the theory of oscillations, and quantum mechanics—a testimony to the power

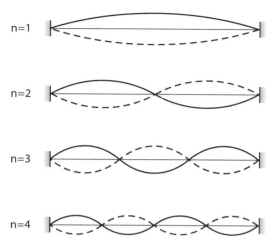

Figure 11.7. The modes of a vibrating string

of mathematics to elevate a once concrete concept to heights of abstraction never imagined by those who first introduced it.

Notes and Sources

1. Hilbert introduced this space in his work in mathematical physics. But it was Erhard Schmidt (1876–1959) and Maurice Fréchet (1878–1973) who gave it a geometric interpretation as a vector space of infinitely many dimensions. See Morris Kline, *Mathematical Thought from Ancient to Modern Times* (New York and Oxford: Oxford University Press, 1990), vol. 3, pp. 1082–1095.

2. For more on Hilbert spaces, see Erwin Kreyszig, *Introductory Functional Analysis with Applications* (New York: Wiley, 1978).

From Flat Space to Curved Spacetime

On June 10, 1854, a new geometry was born.
—Michio Kaku, *Hyperspace*, p. 30

$\mathbf{A}$s already mentioned, the early 1800s saw a renewed interest in both synthetic and projective geometry, leading to the discovery of a host of new properties of ordinary geometric objects like circles and polygons. However, as important as these discoveries were, they did not fundamentally change the nature of mathematics; they were more embellishments upon the foundations laid by Euclid two thousand years earlier. But this changed dramatically around 1830 with the creation of two new branches of mathematics that would profoundly affect its future course: differential geometry and non-Euclidean geometry.

Differential geometry, founded by Euler and brought to its modern form by Gauss and Riemann (see below), applies the methods of calculus to the study of surfaces. Whereas metric properties—those properties of a surface that can be measured and expressed as numerical quantities—have little place in projective geometry, differential geometry is all about metrics; in fact, the very concept of a *metric* is central to its development.

But first we must say a word about coordinates. When Descartes invented coordinate geometry in 1637, he used two intersecting lines as axes (though he did not call them axes, and they were not necessarily perpendicular). Any point in the plane could be located by specifying two numbers, its distances from the two lines (fig. 12.1). This in time evolved into our familiar rectangular coordinate system which, together with polar coordinates, was sufficient to investigate the properties of most plane curves. But to locate a point on the surface of a three-dimensional solid, other coordinate systems may be preferable.

The surface of a sphere provides a good example. The familiar circles of longitude and latitude on a globe form a coordinate system in which the coordinates are not distances, but angular measures. When we say that Jerusalem has longitude 35° east and latitude 32° north, we mean that it is located 35° east of the meridian of zero longitude through Greenwich, and 32° north of the equator. Of course, knowing Earth's radius, we can translate these angular

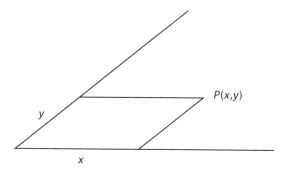

Figure 12.1. Coordinates

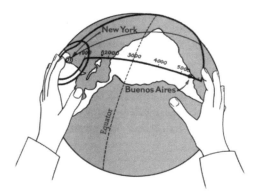

Figure 12.2. A globe and geometer

coordinates into actual distances in space. But this would defy the whole purpose of these coordinates, which is to locate a point not in three-dimensional space, but *on the surface on which it lies.*[1]

Would the Pythagorean theorem hold in such a coordinate system? To find out, I have in front of me a National Geographic globe equipped with a *geometer*, a spherical ruler capable of measuring distances on the surface of the globe (fig. 12.2). Using this device, I find that the straight-line distance between New York and London is about 3,500 statute miles. This "straight line" route is actually an arc of a *great circle*, a circle whose center coincides with the center of the globe. The great-circle arc between two points on a sphere is the shortest distance between them.

Now imagine an aircraft flying from New York to London in two legs: first due east along the circle of latitude of New York (40°N) until the plane reaches the meridian of London (0° longitude), and then due north along this meridian. The first leg is 3,863 miles long, the second 726 miles. The two legs, together

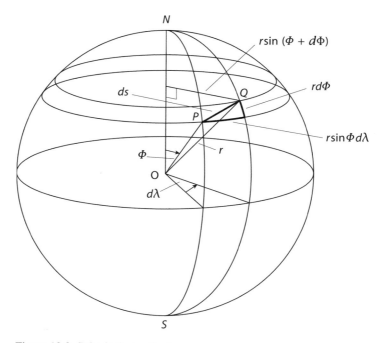

Figure 12.3. Spherical coordinates

with the straight-line route, form the three sides of a spherical right triangle. For this triangle we have $3,863^2 + 726^2 = 15,449,845$, whereas $3,500^2 = 12,250,000$. So $a^2 + b^2 > c^2$. We conclude that *on the surface of a sphere, the Pythagorean theorem does not hold.*

In the case of a sphere, there is a formula that gives the distance between two arbitrary points in terms of their coordinates on the surface; but for a general surface, no such formula exists. The best we can do is to find the distance *ds* between two *infinitesimally close* points. Let us illustrate this for the surface of a sphere of radius r (fig. 12.3). Let the points be $P(\lambda, \phi)$ and $Q(\lambda + d\lambda, \phi + d\phi)$, where λ and ϕ are, respectively, the longitude and latitude of P, each measured in radians instead of the usual degrees. (It is also common in mathematics to measure latitude from the North Pole ($\phi = 0$) to the equator ($\phi = \pi/2$), in reverse of the convention in geography.) The point P lies on a circle of latitude with radius $r \sin \phi$, so the arc length on this circle between the longitudes λ and $\lambda + d\lambda$ is $(r \sin \phi) \, d\lambda$. On the other hand, the arc length along the meridian λ between the latitudes ϕ and $\phi + d\phi$ is $r \, d\phi$. And since these arcs are perpendicular to each other, we have

$$ds^2 = (r^2 \sin^2 \phi) \, d\lambda^2 + r^2 d\phi^2. \tag{1}$$

Comparing this expression with its counterpart in rectangular coordinates, $ds^2 = dx^2 + dy^2$, we notice one important difference (aside from the symbols):

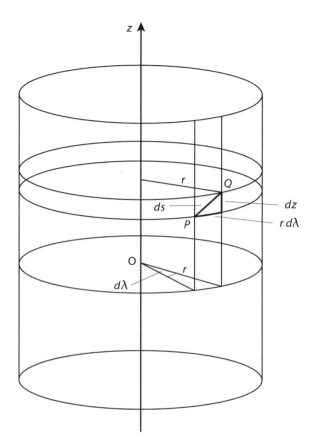

Figure 12.4. Cylindrical coordinates

the coefficient of $d\lambda^2$ is not only different from 1, but is itself a function of ϕ. That is, the arc length depends not only on the *difference* in longitude and latitude between the two points, but also on their *actual* latitude—a consequence of the fact that the circles of latitude shrink in size as we approach the pole. This indicates that the surface of a sphere, *when regarded as a two-dimensional space*, has an intrinsically different geometry than the plane.

Let us consider another example: the surface of a right cylinder (we assume the cylinder extends indefinitely up and down). Let the base be a circle of radius r. As with the sphere, we can enmesh the cylinder in a coordinate system in which lines of longitude ("meridians") run along the cylinder, and circles of latitude ("parallels") perpendicular to it (fig. 12.4). A point P can be located by its longitude λ and its latitude z (its distance from the "equator," measured along the meridian through P). The arc length ds between two neighboring points $P(\lambda, z)$ and Q ($\lambda + d\lambda$, $z + dz$) is given by

$$ds^2 = r^2 d\lambda^2 + dz^2. \tag{2}$$

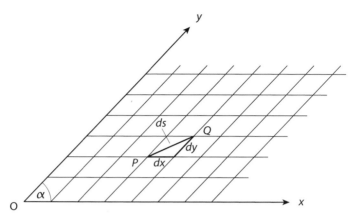

Figure 12.5. Slanted coordinates

Thus, except for the scaling factor r in the first term, the Pythagorean theorem maintains its "natural" form. This is a strong indication that the surface of a cylinder has the same intrinsic geometry as the plane; for example, we can flatten the cylinder onto a sheet of paper without distortion (which is impossible to do with a sphere). The surface of a cylinder follows the laws of Euclidean geometry; it forms a two-dimensional *Euclidean space*.[2]

We see, then, that every coordinate system has its own expression for ds^2, its own form of the Pythagorean theorem. Because of its importance in the study of surfaces, this expression is given the name *metric* (used here as a noun). By looking at the metric, we can tell a lot about the nature of the surface under consideration. Consider, for example, the angle of intersection of the coordinate lines with one another. In the two cases discussed above, the metric contained only "pure" terms; that is, each term involved only a single coordinate differential. Whenever this happens, the grid lines cross each other at right angles; the coordinate system is said to be *orthogonal*. In a nonorthogonal system, the metric contains "mixed" terms involving products of coordinate differentials. To illustrate this, consider a coordinate system in which the x- and y- axes are straight lines inclined to each other at an angle α (fig. 12.5). The grid lines in this system are slanted, each "square" having the shape of a diamond. Using the Law of Cosines, we can write the distance between two points $P(x_1, y_1)$ and $Q(x_2, y_2)$ as

$$s = \sqrt{\Delta x^2 - 2\,\Delta x\,\Delta y \cos(180° - \alpha) + \Delta y^2} = \sqrt{\Delta x^2 + 2\,\Delta x\,\Delta y \cos\alpha + \Delta y^2},$$

where $\Delta x = x_2 - x_1$ and $\Delta y = y_2 - y_1$. The metric for this coordinate system is therefore

$$ds^2 = dx^2 + 2\cos\alpha\, dx\, dy + dy^2. \tag{3}$$

The fact that this system is nonorthogonal is revealed by the presence of the mixed term $2 \cos \alpha \, dx \, dy$.

The development of these ideas was by and large complete by 1850, when a young mathematician came to the scene and gave them an entirely new meaning. Georg Friedrich Bernhard Riemann was born in 1826 in Hannover, Germany, to a Lutheran minister and his wife. His early interest in theology got him involved in an attempt to prove the truth of the Book of Genesis by mathematical methods. Soon, however, Riemann's attention turned to more fruitful areas of mathematical research, foremost among which was geometry. In his 1851 doctoral dissertation under the guidance of Gauss, he introduced two groundbreaking ideas: that geometry need not be limited to three dimensions, and that the properties of space—specifically its metric—may change from one point to another. The metric, in other words, is a *local* rather than global quantity.

Today the fourth dimension has become part of our daily language, and even the talk of ten dimensions no longer arouses a sense of dismay. But in the nineteenth century the idea of higher dimensions was utterly novel, the stuff of science fiction if not of total nonsense. After all, was it not obvious that a point has zero dimension, a line has one, a plane two, space three—and that was the end of it? In his dissertation, the twenty-five-year-old Riemann introduced four-dimensional space—and in fact spaces of *any* number of dimensions—as a mathematical reality. Whether such spaces have a *physical* reality was beside the point. Mathematics, Riemann insisted, can be allowed to create its own spaces, as long as they were free of internal contradictions.

But that was not all. Every such space, Riemann argued, has its own metric, given by the differential equation

$$ds^2 = \sum_{ij} a_{ij} dx_i dx_j. \tag{4}$$

The sigma sign in this equation is actually a double sum: it tells us to form all possible products $dx_i \, dx_j$, each multiplied by a coefficient a_{ij}, and then sum these products for all i and j. And because it would be redundant to count the term $dx_i \, dx_j$ twice, once as $dx_i \, dx_j$ and again as $dx_j \, dx_i$, we adopt the convention that $a_{ij} = a_{ji}$. For example, in the case of the slanted coordinate system (equation (3)) we have $a_{11} = 1$, $a_{12} = a_{21} = \cos \alpha$, $a_{22} = 1$.

The coefficients in equation (3) were constants, because the angle α was fixed. This means that the slanted coordinate system has the same metric everywhere. In the general case, however, the coefficients a_{ij} are allowed to be functions of the coordinates $x_1, x_2, \ldots, x_n$. We already saw an example of this in the surface of a sphere (equation (1)), where the coefficient ($r^2 \sin^2 \phi$) of $d\lambda^2$

is a function of the latitude ϕ. Thus the metric, in general, is a *local* rather than global property of the coordinate system we choose to describe a space.

Often in this book I have drawn parallels between mathematics and music, and the present discussion suggests a particularly apt case. The metric is to geometric space what the *meter* is to musical space: each defines the fabric on which the subject matter is woven. In classical music—roughly up to 1800—a movement usually had a fixed meter: 4/4 (meaning four beats to a measure), 3/4, 6/8, and occasionally 5/4. The meter set the character of the work just as much as the melodic theme, and sometimes even more so. The five drum beats that open Beethoven's Violin Concerto, barely audible at first but repeated again and again as the work progresses, imprint their rhythmic pattern on the entire first movement. But by the mid-nineteenth century—about the time of Riemann's pioneering work—composers began using meters that frequently changed within a single movement. In Igor Stravinsky's music for the ballet *The Rite of Spring*, the rhythm routinely varies from bar to bar, at one place changing from the predominant 6/8 to 7/8, then to 3/4, 6/8, 2/4, 6/8, 3/4, and finally to 9/8, only to be followed by a second succession of alternating meters. No wonder the audience at the work's premiere performance in Paris in 1913 reacted negatively: at the first sounding of the strange, discordant notes, pandemonium broke out, with people whistling, stomping their feet, and honking auto horns. The comfortable old world of flat space and fixed rhythm may be gone for good.

The two ideas Riemann introduced—the existence of spaces of an arbitrary number of dimensions and the notion of variable geometry—forced mathematicians to rethink the way they looked at a surface. The concept of a curved surface, of course, did not originate with Riemann. But prior to him, a surface had always been regarded as the envelope of a three-dimensional solid, to be studied as part of the geometry of that solid. A good example of this is spherical trigonometry, a subject of great importance in navigation. The laws of spherical trigonometry are markedly different from those of plane trigonometry, but this surprised no one; spherical trigonometry simply applied the geometry of a sphere to its own surface.

Riemann at once changed this view. He regarded a surface as a space in its own right, detached from any solid of which it may be a boundary. Such a space has its own geometry, which may be different from the Euclidean geometry of flat space. Imagine an antlike, two-dimensional creature that knows only two kinds of movement, forward and backward, and left and right. The third dimension does not exist in this ant's life; for all intents and purposes, it spends its years in a two-dimensional, flat world. But if this ant would start exploring the world in which it lives—the surface of our spherical planet—it would discover that this world does not follow the good old Euclidean rules.

For example, if two ants set out simultaneously from two distinct points on the equator, each heading due north, there would be no doubt whatsoever in their minds that they are moving along parallel lines. Imagine their surprise when, after eons of travel, they meet at the North Pole! The seemingly parallel lines turned out to converge: from our three-dimensional vantage point, they are not parallel at all. But our ants don't know this; from their two-dimensional perspective, parallel lines do indeed meet.[3]

Other familiar laws of Euclidean geometry also fail on the surface of a sphere: a straight line (actually an arc of a great circle) does not extend indefinitely; it has a finite length equal to the circumference of the sphere. The sum of the angles of a triangle is always greater than 180°, and the Pythagorean theorem does not hold, at least not in the form $a^2 + b^2 = c^2$. Again, all these facts had been known long before Riemann, but it was Riemann's new outlook that changed the way we interpret them. To use his mode of expression, the surface of a sphere is a *two-dimensional non-Euclidean space.*

One often hears the phrase "curved space," usually as a contrast to Euclidean flat space. But the words "flat" and "curved" must be taken with a grain of salt. The curved surface of a cylinder, for example, when regarded as a space in its own right, is "flat" in the sense that it follows the laws of Euclidean geometry; in particular, it has a Pythagorean metric (see equation (2)). By contrast, the surface of a sphere is "curved" even when regarded as a two-dimensional space, independently of the solid sphere beneath it. This is reflected in the non-Pythagorean form of equation (1).[4]

But suppose now our ants inhabit a truly (from our three-dimensional perspective) flat space, a plane, but a plane with occasional wrinkles. For us outsiders, these wrinkles are no more than local intrusions of the two-dimensional plane into the third dimension. But from the ants' point of view, these wrinkles are local disturbances in the overall flatness of the plane— momentary changes in the geometry of their world. In Riemann's view, we must abandon the notion that a space has a fixed, preordained geometry. On the contrary, he regarded all geometric properties, and in particular the metric, as *local* properties that vary from point to point. This is why the coefficients a_{ij} in the expression $\Sigma_{ij}\, a_{ij}\, dx_i\, dx_j$ are allowed to be functions of the coordinates: each point has its own geometry, its own metric, its own form of the Pythagorean theorem. Geometry, according to Riemann, is local. And this is true for spaces of any number of dimensions.

On June 10, 1854, Riemann gave a historic address before the mathematics faculty of the University of Göttingen. It was his habilitation speech, a long-honored rite of passage that every aspiring scholar had to go through to be inducted into the hallowed halls of academia. Among the distinguished professors attending was his mentor Gauss, seventy-seven years old and in his last year of life. Riemann's address was entitled "On the Hypotheses Which Lie at the Foundation of Geometry." Notwithstanding the revolutionary nature of his ideas, his talk was met with great enthusiasm, and he had every reason to look forward to a long, productive career at the frontier of mathematical research. But it was not to be. Suffering from poor health since childhood, he died of tuberculosis on July 20, 1866, two months short of his fortieth birthday. His speech was published posthumously in 1868; fifty years later his ideas would be the cornerstone of the general theory of relativity.

Notes and Sources

1. I can't resist telling here a story I once heard from a software specialist who was assigned with checking the accuracy of a new navigational system that had just been installed on an aircraft. The test run called for a flight from England due north over the North Pole and on to Alaska. As the plane approached the pole, the readout displayed the latitude: 88°, 89°, 90°, 91°, ...

2. However, had we used the geographical latitude ϕ (measured at the center of the equator) instead of the linear latitude z, the expression for the arc length would be $ds^2 = r^2[(d\lambda)^2 + \sec^4 \phi \, (d\phi)^2]$, which has a distinctly non-Euclidean look. This shows that the metric is not the only factor determining the nature of the surface; of equal importance is the *curvature function* on the surface.

3. To some extent, this is really a semantic issue. The last of the twenty-three definitions opening Euclid's *Elements* defines parallel lines as "straight lines which, being in the same plane and being produced indefinitely in both directions, do not meet one another in either direction." Since the paths of the two ants along their respective meridians converge at the North Pole, they are not parallel.

One might be tempted to regard the circles of latitude as parallel "lines." True, these circles do not meet one another (which is why they are called "parallels" in geography); but except for the equator, none of them is a great circle, and so they do not qualify to be called "straight lines" on the sphere.

4. A beautiful example of a curved space is provided by the *Cloud Gate*, a huge reflecting sculpture by the Indian-born British artist Anish Kapoor (b. 1954) at the Millennium Park in Chicago. Known popularly as the "Bean" (see plate 3), it transforms the rectangular grid of the plaza around it into a curved grid on its shiny surface.

A Case of Misuse

Let no one dare to attribute the shame
of misuse of projections to Mercator's name;
but smother quiet, and let infamy light
upon those who do misuse, publish or recite.
—An anonymous writer, ca. 1600

In 1990 I took what was to be my last flight with TWA, once the pride of American aviation but now sadly defunct. Having nothing else to do, I picked up the shopping catalog from the pocket at the back of the seat in front of me and soon came across a photo that caught my attention (fig. S9.1). It featured a smiling lady who seems to be measuring the distance between two locations on a large map. Nothing unusual about that, except that the map happens to be Mercator's map, famous (or perhaps infamous) for its glaring distortions of the size of continents. Even a cursory comparison with a globe will show that Greenland is much too large on this map—larger in fact than South America, although in reality it is nine times smaller. Can it be that such a distorted representation of our planet has become the most famous map in the history of cartography?

This takes us back to the sixteenth century, the century of exploration and discovery. Mariners were venturing ever farther into unknown seas, driven by visions of new lands and fabulous riches. But two major obstacles hampered their quest: no reliable method was known for determining a ship's longitude at sea, and no map existed on which a *rhumb line*—the compass bearing that directs a navigator from his port of departure to his destination—showed as a straight line. The first problem would not be settled for another 150 years;[1] the second was solved in 1569 by the Dutch cartographer Gerhard (or Gerardus) Mercator (1512–1594).[2]

As everyone who has tried to press the peels of an orange against a table knows, it is impossible to flatten the surface of a sphere onto a sheet of paper without causing significant distortions. To cope with this

Figure S9.1. Mercator's map

problem, cartographers have invented a variety of *map projections*—mathematical functions that assign to each point on the globe a unique "image" point on the map. There exist numerous projections that accomplish this, each with its own distortions, but also with its *invariants*—features that do not change under the projection. Mercator's goal was to find a projection that would preserve *direction*. Such a map would show all rhumb lines—lines of constant bearing on the globe—as straight lines, making it easy for a navigator to plot his course on the map and follow it at sea.

To accomplish this, Mercator chose a rectangular grid in which all meridians (circles of longitude on the globe) showed as vertical, equally spaced straight lines. The parallels (circles of latitude) were horizontal lines of equal length, but their spacing gradually increased with latitude (fig. S9.2). This was necessary in order to compensate for the shortening of the circles of latitude as one approached either pole. Figure S9.3 shows a globe of radius R and a circle of latitude ϕ on it. The true circumference of this circle is $2\pi R \cos \phi$, but on the map its length is constant and equal to $2\pi R$. Thus the length of each parallel on the map is stretched by a factor $2\pi R/(2\pi R \cos \phi) = \sec \phi$ relative to its true length. Mercator realized that in order to preserve direction, the spacing *between* neighboring parallels must also be increased by the same factor. This leads to the differential equation

$$dy = R \sec \phi \, d\phi. \tag{1}$$

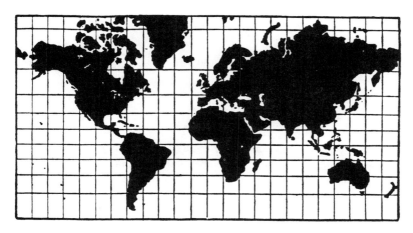

Figure S9.2. The Mercator grid

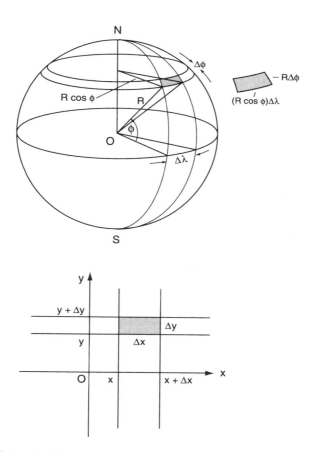

Figure S9.3. A spherical rectangle and its image on Mercator's map

Here the latitude ϕ is measured in radians, rather than in the usual degrees, minutes, and seconds. The meridians, on the other hand, are everywhere equally spaced, and so

$$dx = Rd\lambda, \tag{2}$$

where λ denotes longitude (again in radians). Mercator, who was not a mathematician, never wrote down these equations explicitly; essentially he solved them by numerical integration (cartographers are still debating how exactly he handled the calculations). It was only in 1599, five years after his death, that Edward Wright (ca. 1560–1615), an English mathematician and instrument maker, provided an explanation of the principles underlying Mercator's map.[3]

As always in cartography, the preservation of one feature—in this case direction—comes at the expense of others. In particular, the Mercator projection distorts the shape of countries at high latitudes. Consequently, the distance between two points on the map is *not* preserved—it is not proportional to the actual distance between them on the globe. And this in turn means that the Pythagorean theorem does not hold true. Indeed, it follows from equations (1) and (2) that the element of arc length ds between two neighboring points on Mercator's map is given by

$$ds^2 = R^2(d\lambda^2 + \sec^2 \phi \, d\phi^2),$$

a distinctly non-Pythagorean formula.

This brings us back to the anonymous quotation at the head of this sidebar. The Mercator map was invented with a single goal in mind: to preserve direction. That it has been used for other purposes is unfortunate, and has caused much confusion and misunderstanding. But that is the fault of those who misused the map, not of its inventor.

Notes and Sources

1. See Dava Sobel, *Longitude: The True Story of a Lone Genius Who Solved the Greatest Scientific Problem of His Time* (New York: Walker, 1995), and *The Illustrated Longitude* (with William J. H. Andrewes; New York: Walker, 1998).

2. On Mercator's life, see Nicholas Crane, *Mercator: The Man Who Mapped the Planet* (London: Weidenfield and Nicholson, 2002), and Andrew Taylor, *The World of Gerard Mercator: The Mapmaker Who Revolutionized Geography* (New York: Walker, 2004).

3. For a more detailed discussion of Mercator's map, see *Trigonometric Delights*, chap. 13. See also Mark Monmonier, *Rhumb Lines and Map Wars: A Social History of the Mercator Projection* (Chicago and London: University of Chicago Press, 2004).

Prelude to Relativity

The famous "Michelson-Morley experiment" was
performed in Cleveland in the year 1887, with an
absolutely negative result. . . . The expected
addition and subtraction of the earth's velocity to
the velocity of light simply did not take place.
  —Cornelius Lanczos, *Albert Einstein and the
  Cosmic World Order*, p. 38

The small, quiet town of Basel is perched among scenic hills on the banks of
the Rhine, where the borders of Switzerland, France, and Germany meet. For
centuries it has been a center of arts and crafts, of printing and publishing, and
of first-rate scholarship. Among its residents were the humanist Erasmus,
the painter Hans Holbein the Younger, and the statesman Theodor Herzl. Its
university, one of the oldest in Europe, (founded in 1460), was home to the
Bernoulli dynasty of mathematicians, as well as to Leonhard Euler. It is here,
in the cloisters of the impressive Münster cathedral overlooking the town, that
Jakob Bernoulli, the first of the dynasty to achieve prominence, is buried. Engraved on the headstone is his favorite curve, the spiral, with the inscription
Eadem mutata resurgo ("though changed, I shall arise the same"), a reference
to the invariance of the logarithmic spiral under rotation, stretching, and inver-
sion. Alas, whether by mistake or intentionally, the stonecutter engraved the
wrong curve—an Archimedean (linear) spiral instead of a logarithmic! No
doubt Jakob would have turned in his grave (see fig. 13.1).

Basel is also a geographic landmark: it is here that the Rhine makes a sharp
turn to the north, changing from a winding, narrow Alpine stream to a mighty
waterway that flows for 500 miles through the heartland of western Europe before it empties into the North Sea near Rotterdam. It was at this spot, on the
left bank of the Rhine, that I first grasped the meaning of relativity.

Like every high school student for the past hundred years, I had to master the
subject of vector addition and the parallelogram rule (see p. 159) in my science
class. It was all nice and clear on the blackboard, but the physical reality of it
eluded me until, some years later, I visited Basel and watched one of the small

Figure 13.1. Jakob Bernoulli's tombstone, Basel

ferryboats that regularly cross the Rhine from bank to bank (fig. 13.2). For about five Swiss francs you can enjoy a leisurely, ten-minute river crossing in a small boat propelled entirely by the stream; there are no engines, no sails, no oars. How does it work? The vessel is loosely tied to a fixed steel cable running overhead across the river. As the swift current takes hold of the boat, the cable forces it to move in a direction perpendicular to the river, seemingly without any forward motion of the boat (fig. 13.3).

Figure 13.2. A ferryboat crossing the Rhine. Note the steep angle to the stream.

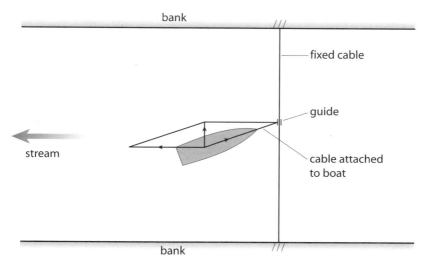

Figure 13.3. What makes the boat cross the river?

As I watched, transfixed by the sight, I noticed that the boat was aimed at a steep angle relative to the stream; in fact, it almost pointed into it. Then I realized that the boat *was* in fact moving forward with respect to the stream, but the force of the stream also carried it backward; only the boat's orientation relative to the stream prevented the two motions from canceling out: it provided the lateral force needed to cross the river.

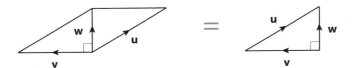

Figure 13.4. Vector addition

To put all this into vector language, let **v** and **u** denote, respectively, the velocity vectors of the stream and the boat's motion relative to it (fig. 13.4). Their resultant, or vector sum, is a vector **w** perpendicular to the stream. We have **w** = **u** + **v**, with $w^2 = u^2 - v^2$ (here u, v, and w denote the magnitude, or speed, of the respective velocity vectors).[1]

This led me to a little "thought experiment," the kind of mental rumination that young Einstein enjoyed. Suppose two identical boats, now fitted with engines, leave simultaneously from a point A on the bank of a river one mile across. One boat crosses the river to a point B directly opposite from A, then turns around and crosses back to A. The other boat travels one mile *along* the river to a point C, then returns to A. Which boat will arrive back at A first?

As before, let the current have speed v, and let each boat move with speed u in still water (we assume $u > v$, for otherwise the second boat will never be able to make the return trip). Using the same notation as before, we have (fig. 13.5) **w** = **u** + **v**, and so $w^2 = u^2 - v^2$. The round-trip time it takes to cross the river is

$$t_1 = \frac{2}{\sqrt{u^2 - v^2}} \tag{1}$$

(note that the situation is entirely symmetric for the trip from A to B as from B to A). The round-trip time along the river is

$$t_2 = \frac{1}{u+v} + \frac{1}{u-v} = \frac{2u}{u^2 - v^2}. \tag{2}$$

To see which of the two is smaller, we rewrite t_2 as follows:

$$t_2 = \frac{2}{\sqrt{u^2 - v^2}} \cdot \frac{u}{\sqrt{u^2 - v^2}}.$$

The first factor is equal to t_1, and the second factor is always greater than 1 (the denominator being less than $\sqrt{u^2} = u$). Thus $t_2 > t_1$; the boat crossing the river will return to the starting point first.

Now let us imagine that each boat is replaced by a ray of light. In the nineteenth century, scientists believed that electromagnetic waves, light included, must travel through a material medium in much the same way as sound waves travel through the air. However, the laws of electromagnetism, summarized in four differential equations known as Maxwell's equations, do not require the

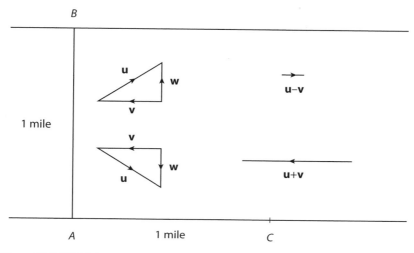

Figure 13.5. Which boat returns to point A first? The upper vector diagrams show the outbound trips; the lower diagrams show the return trips.

presence of a material medium; the "medium" is the electromagnetic field itself. Yet nineteenth-century physics was still dominated by Newton's mechanistic view of the universe, so physicists invented the idea of *ether*, an invisible, "luminiferous" substance that was supposed to fill all of space and allow light to pass through. The elusive ether served the additional purpose of providing an absolute, universal frame of reference, itself at rest, relative to which all motion was to be measured.

The ether hypothesis became a fixture of nineteenth-century physics, yet no one found any empirical evidence that it really existed. In an attempt to settle the issue, two American scientists, Albert Abraham Michelson (1852–1931) and Edward Williams Morley (1838–1923), in 1887 conducted an experiment in which a beam of light was split by a semitransparent mirror, aimed at 45° to the direction of the beam, into two separate beams traveling at right angles to each other (fig. 13.6). These in turn were reflected back along their paths by two secondary mirrors, positioned at equal distances from the splitting mirror, where they were recombined. The entire apparatus was oriented so that one beam traveled in the direction of Earth's motion around the Sun, while the other traveled perpendicularly to it.

In the absence of the ether, the two beams would arrive back at their source at exactly the same instant (indeed, going back to equations (1) and (2), if $v = 0$, then $t_1 = t_2$). But if the ether existed, as everyone had assumed, the beam traveling parallel to Earth's motion will return to its starting point a fraction of a second later than its perpendicular counterpart; and because each beam is a wave, the tiny difference in their time of arrival would cause the two waves to arrive out of phase. By adjusting the position of the secondary mirrors, one could force

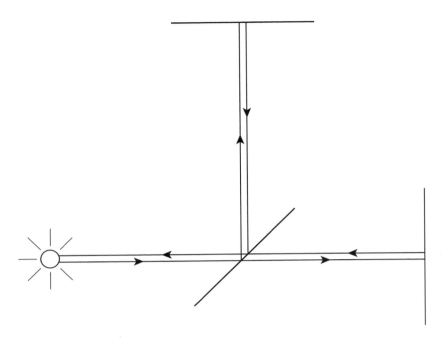

Figure 13.6. The Michelson-Morley experiment

the two waves to recombine exactly in antiphase—the crest of one wave coinciding with the trough of the other. A series of interference fringes—alternating dark and bright lines—would be formed and could be observed on a screen.[2]

To Michelson and Morley's surprise, no interference fringes were ever observed, even after repeated trials. This at once caused ripples in the scientific community. Various attempts were made to explain the negative result, none of them convincing. It was a twenty-six-year-old physicist who finally gave the correct explanation: the experiment "failed" because there was no ether, period. The ether was fiction, and so was absolute rest. The physicist was Albert Einstein, working as a clerk at the Swiss Patent Office in Bern and still unknown to the world. His groundbreaking paper, "On the Electrodynamics of Moving Bodies," appeared in 1905 in the *Annalen der Physik* and at once shook the foundations of physics; it would forever change the way we perceive space and time. The theory of relativity was born.

At the end of his article, Einstein thanked his friend Michele Besso "for several valuable suggestions" in developing his theory. Not mentioned was Einstein's indebtedness to a scientist who lived nearly 2,500 years before him: Pythagoras. For the footprints of the Pythagorean theorem appear everywhere in the theory of relativity, from the omnipresent $\sqrt{1 - v^2/c^2}$ of the special theory to the enigmatic $ds^2 = \sum_{ij} g_{ij}\, dx_i\, dx_j$ of general relativity, the subject of our next chapter.

Notes and Sources

1. The last equation, despite its seeming disagreement with the vector equation preceding it, is in fact consistent with it; to convince yourself of this, look again at figure 13.4.

2. This description of the Michelson-Morley experiment is, of course, greatly simplified. For a more complete discussion, see David Layzer, *Constructing the Universe* (New York: Scientific American Library, 1984), pp. 157–160.

From Bern to Berlin, 1905–1915

Space by itself and time by itself must sink into the
shadows, while only a union of the two preserves
independence.
  —Hermann Minkowski, "Space and Time," 1908

To reconcile the propagation of electromagnetic waves with the known
facts, Einstein started with three postulates:

1. All motion is relative. *An absolute state of rest is fiction.*
2. The laws of physics are the same for all observers, regardless of their
   motion relative to one another.
3. Light propagates through space with the same speed, regardless of
   the observer's own motion relative to it; that is, *the speed of light in
   vacuum is the same for all observers.*

In the closing years of the nineteenth century, well before Einstein's name was
known, a lively debate on these questions was already taking place in the
physics community, and the negative results of the Michelson-Morley experiment (see p. 186) only added urgency to it. But it was Einstein who drew the
correct conclusions from the facts. He realized that postulate 3 implies that
time must be relative as well. Whereas Newton had assumed the existence of a
kind of universal master clock that marks the passage of time uniformly
throughout space, Einstein realized that each observer experiences his or her
own time.[1]

  Imagine now a source of light located at the origin of a three-dimensional
coordinate system (x, y, z). At time $t = 0$ the source emits a spherical wave that
propagates through space at the speed of light c. Let $P(x, y, z)$ be a point on the
wave front of the propagating wave. After t seconds, P has traveled a distance
ct, so we have

$$x^2 + y^2 + z^2 = c^2t^2. \tag{1}$$

This is the equation of the wave front as seen by an observer at the origin of
the (x, y, z) system. Now imagine a second observer, situated at the origin of a

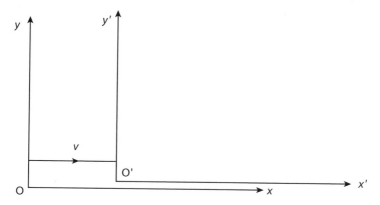

Figure 14.1. Two frames of reference in relative motion

coordinate system (x', y', z') that is moving relative to the (x, y, z) system with a constant speed v. According to postulate 3, the second observer, too, should see the same wave front propagate at the speed c. But this means that

$$x^2 + y^2 + z^2 - c^2 t^2 = x'^2 + y'^2 + z'^2 - c^2 t'^2. \qquad (2)$$

Note that all the variables on the right side of equation (2) are primed, but c is not: the speed of light is the same in the two systems.

From equation (2) one can derive the transformation equations from the (x, y, z, t) system to the (x', y', z', t') system (note that we have now included time as a fourth coordinate, a precursor of things to come). To simplify matters, we choose our coordinate system so that the primed system moves along the positive x-axis of the unprimed system (fig. 14.1); we then have $y' = y$, $z' = z$. I will spare the reader the details of the derivation, which are well known and can be found in any textbook on modern physics.[2] The result is

$$x' = \frac{x - vt}{\sqrt{1 - v^2/c^2}}, \quad y' = y, \quad z' = z, \quad t' = \frac{t - (v/c^2)x}{\sqrt{1 - v^2/c^2}}. \qquad (3)$$

Taken together, equations (3) are the famous Lorentz transformation, named after the Dutch physicist Hendrik Antoon Lorentz (1853–1928), Einstein's senior colleague and trusted friend. The expression $\sqrt{1 - v^2/c^2}$ is at the core of special relativity; it appears in nearly every relativistic formula, including the most famous of them all: $E = \frac{m_0}{\sqrt{1 - v^2/c^2}} c^2$, commonly known as $E = mc^2$.[3]

❖   ❖   ❖

Einstein published his theory in 1905 in volume 17 of the respected journal *Annalen der Physik* (the same volume carried two more articles by him, each groundbreaking in its own right). Reaction was at first slow to come. As an outsider—he did not even have an academic position at the time—Einstein was hardly known to professional physicists. But a few scientists did take notice. Among the first to embrace the new theory was one of Einstein's former professors at the Swiss Polytechnic in Zurich, Hermann Minkowski (1864–1909). Born in Russia, Minkowski went to Königsberg (then in Germany and now renamed Kaliningrad, Russia), where he got his Ph.D. in mathematics. He later moved to Zurich and still later to Göttingen, the world-renowned center of mathematical research. Minkowski's opinion of young Einstein, his student at the Polytechnic, was at first rather low, a view he would soon come to regret. When Einstein published his paper on relativity, Minkowski became an instant convert.

Looking at equation (2), Minkowski was struck by the similarity of either side to the Pythagorean theorem. In a three-dimensional coordinate system, the length d of the radius vector from the origin to the point (x, y, z) is given by $d^2 = x^2 + y^2 + z^2$. Imagine now that this coordinate system is undergoing a rotation in space about the origin, which itself stays fixed. The point (x, y, z) will then have new coordinates (x', y', z'), which can be found from the old coordinates (x, y, z) if the angle of rotation and the direction of its axis are known. But there is one quantity that does *not* change in the process: the length of the radius vector. That is to say, $x^2 + y^2 + z^2 = x'^2 + y'^2 + z'^2$. The expression $x^2 + y^2 + z^2$, in other words, is *invariant* under rotation.

Minkowski immediately saw the similarity between this kind of rotation and equation (2). But there were two glitches. For one, we are dealing here with *four* variables, so if there is any merit to the comparison, we will have to work with a four-dimensional coordinate system (x, y, z, t). However, Riemann had already legitimized the fourth dimension some fifty years earlier, so this was not really an issue. The second glitch was more serious: the fourth term on each side of the equation, namely $c^2 t^2$ and $c^2 t'^2$, each comes with a *negative sign*—definitely not the form of the Pythagorean theorem.

It was at this juncture that Minkowski had a flash of inspiration. Since, he reasoned, we are already dealing with a four-dimensional space—once thought of as a purely imaginary construct but by by now fully accepted as a mathematical reality—why not define a new, *imaginary coordinate*, where the word "imaginary" now refers to $i = \sqrt{-1}$? Minkowski therefore introduced a new variable, $m = ict$. Since $m^2 = (ict)^2 = -c^2 t^2$, equation (2) takes the neat form

$$x^2 + y^2 + z^2 + m^2 = x'^2 + y'^2 + z'^2 + m'^2. \tag{4}$$

That is to say, the quantity $x^2 + y^2 + z^2 + m^2$ is invariant under the Lorentz transformation. Thus, in Minkowski's interpretation, the transformation from the (x, y, z, t) to the (x', y', z', t') system amounts to a rotation in the four-dimensional space (x, y, z, m).

But it was more than just a mathematical device. In his lecture of 1908 entitled "Space and Time," Minkowski declared that the two can no longer be regarded as separate entities; they must be thought of as one, inseparable reality, henceforth to be called *spacetime*. By giving the fourth dimension a physical reality, Minkowski inaugurated it into our popular culture. Four-dimensional spacetime changed our perception of the world; it changed our mode of thinking, permeated our language, and even made its way into art in the form of cubism and surrealism. Minkowski proceeded to develop his idea into a powerful mathematical tool, extending the methods of three-dimensional vector algebra and vector calculus to four-dimensional spacetime. In Minkowski's interpretation, every event in spacetime defines a vector with components (x,y,z,m), to be called a *world vector*. To illustrate, when two people agree to meet on a given day at 7 P.M. at the intersection of Fifth Avenue and 42nd Street in Manhattan, that meeting defines an event with coordinates $(5,42,0,7ic)$, where c is the speed of light (the coordinates refer to the grid map of New York City; $z = 0$ indicates that Manhattan is at sea level, and 7 is the hour on the particular day of the meeting).

The length, or magnitude, of a world vector is the quantity

$$\sqrt{x^2 + y^2 + z^2 + m^2} = \sqrt{x^2 + y^2 + z^2 - c^2 t^2}. \tag{5}$$

Two distinct events define an *interval in spacetime*, whose length is given by the distance formula

$$d = \sqrt{(x_2 - x_1)^2 + (y_2 - y_1)^2 + (z_2 - z_1)^2 + (m_2 - m_1)^2}$$
$$= \sqrt{(x_2 - x_1)^2 + (y_2 - y_1)^2 + (z_2 - z_1)^2 - c^2 (t_2 - t_1)^2}. \tag{6}$$

Just as we can find the distance between two points in space by using the distance formula from analytic geometry, so does equation (6) allow us to find the distance between two events in spacetime—their separation in both space and time. To use again the example above, if the two friends agree on a second meeting, to take place on the same day at 10 P.M. at the intersection of Third Avenue and 34th Street, the two events will be separated by a distance

$$\sqrt{(3 - 5)^2 + (34 - 42)^2 + (0 - 0)^2 - c^2 (10 - 7)^2} = \sqrt{68 - 9c^2}.$$

The effect of Minkowski's work was to change the nature of special relativity from a rather simple theory, based on physical intuition, into a highly abstract mathematical subject. Einstein, who was always seeking simplicity in his work, reportedly complained, "Since the mathematicians have attacked the relativity theory, I myself no longer understand it."[4] But he was soon to find in Minkowski's formulation the very mathematical tool he needed to develop his general theory of relativity.

Sadly, Minkowski was not to witness the revolution his ideas would bring about. On January 12, 1909, he died of peritonitis, not yet forty-four years old. On his deathbed he lamented, "What a pity that I have to die in the age of relativity's development."[5]

The story of general relativity has been told in so many books, popular as well as scholarly, that it would be superfluous to repeat it here in detail. Briefly, Einstein was not satisfied with his special theory of relativity ("special" because it applied only for the special case of two observers moving relative to each other with constant speed). His dissatisfaction stemmed from his rejection of the Newtonian concept of "action at a distance," a kind of invisible hand that was supposed to transmit the Sun's gravitational pull on Earth instantaneously over a distance of some 150,000,000 kilometers.[6] The concept could not be supported by experiment, and in any case it ran smack against one of special relativity's fundamental tenets, that nothing in the universe, including gravity, can travel faster than light. Having laid to rest the ether, with its implied assumption of a universal frame of reference at absolute rest, Einstein in 1907 set out to dethrone the idea of action at a distance as well. To do so, he resorted to geometry.

In Einstein's view, a ray of light, being the fastest means to transmit a signal through space, must always travel along the shortest possible path. In the absence of matter, this path is a straight line in spacetime. But put a material object in its way, and a light ray will be bent—not because of any force acting on it, but because the presence of the object causes spacetime to curve. The situation is often compared to a smooth ball placed on a taught, flexible membrane. If the ball is taken off, the membrane will be perfectly flat. But place the ball on it, and the membrane will curve, forcing the ball to move along a path determined by its own weight. As the physicist John Archibald Wheeler (1911–) put it, "Matter tells space how to curve, and space tells matter how to move."

To develop his idea into a full-fledged theory, Einstein needed one mathematical tool with which he was not quite familiar at the time: Riemann's differential geometry of curved, *n*-dimensional space. We all heard the story—quite exaggerated and with little factual basis—that young Einstein was poor in school mathematics (as we saw on p. 117, at a young age he had proved the Pythagorean theorem on his own—not exactly characteristic of someone weak in mathematics). It is true, however, that early in his scientific career Einstein believed that simple mathematics would suffice for his quest to challenge classical, Newtonian physics. He would soon realize how naive that belief had been.

Riemann's geometry led to the creation of a new mathematical concept, a *tensor*, a generalization of the familiar vector of *n* dimensions to a construct of $m \times n$ components subject to certain rules. Tensor analysis (also known as absolute differential calculus) was developed chiefly by two Italian mathematicians, Gregorio Ricci-Curbastro (1853–1925) and his student Tullio Levi-Civita (1873–1941). It is a highly abstract subject, and any attempt to explain it in nontechnical terms will necessarily be inadequate. When Einstein started working on his general theory of relativity, he lacked a fluent knowledge of this subject. In desperation, he turned to a longtime friend from his student days in Zurich, Marcel Grossmann, who taught Einstein the rudiments of tensor analysis.

Equipped with the necessary mathematical tools, Einstein—now director of the venerated Kaiser Wilhelm Institute of Physics in Berlin—was able to finish his theory late in 1915, publishing it the following year. It was hailed as the most elegant physical theory ever proposed, and it was soon confirmed during the total solar eclipse of May 29, 1919, at which the bending of starlight passing near the Sun was measured and found to agree with Einstein's prediction.[7] The results were announced at a special meeting of the Royal Society in London on November 6 of that year, and overnight Einstein became world famous.

At the core of general relativity is the *principle of equivalence*, said to have occurred to Einstein while he tried to imagine a person falling from a roof (and surviving to tell about it). That person, Einstein realized, would not experience any gravity at all: he would be weightless. But suppose the same person were enclosed in an elevator in space, far away from Earth and from any other gravitational influences. If the elevator would be pulled up at precisely the acceleration of free fall (about 9.81 meters/sec^2), the person inside would, for all intents and purposes, think he is being pressed to the floor by the force of gravity: his weight would be the same as if he were standing on solid ground back on Earth. This "thought experiment," Einstein's favorite mode of argument, convinced him that there is no difference between acceleration and gravity: the two are one and the same.

Today, when images of astronauts floating weightlessly in their spacecraft are a household feature, the principle of equivalence is no more a mystery. But in 1907, when Einstein started thinking about the nature of gravity, this idea was far from obvious. Air travel was still in its infancy, and space flight the stuff of science fiction; the fastest speed a person could experience was a fast-moving passenger train (indeed, many of the early popular accounts of relativity used trains for the purpose). So it took a while for the principle of equivalence to gain acceptance. Far more difficult, though, was the mathematics Einstein used in formulating his theory—tensor algebra and calculus. The story goes that when general relativity was published, only three scientists in the world could understand it—one of them Einstein himself. When Sir Arthur Eddington (1882–1944), the eminent English astrophysicist and an early convert to relativity (it was he who organized the 1919 eclipse expeditions) was told this story, he quipped, "Who is the third?"

Einstein now contemplated a second thought experiment. Suppose the elevator moves up at constant speed relative to some reference system. Let a beam of light from a remote source, say a star, enter the elevator through a small slit in one wall (fig. 14.2a). By the time the beam hits the opposite wall, the elevator has moved up a tiny distance, so the beam will hit the opposite wall at a point a little lower than the slit. The path of the beam will still be a

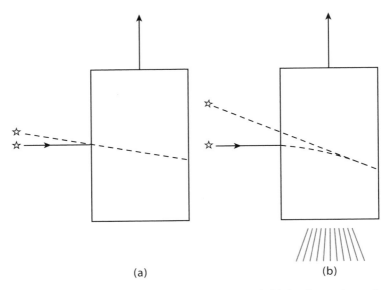

Figure 14.2. (a) An elevator moving up at a constant speed; (b) the elevator is accelerated upward.

straight line, but it will be slightly deflected from its original direction. The person riding in the elevator, not being aware of his own motion, would interpret this small shift as due to a slight change in the direction of the source of light. This is the phenomenon of *aberration*, known since the eighteenth century and apparent as a tiny annual shift in the position of the fixed stars due to Earth's motion around the Sun; as such it has nothing to do with relativity.

But now suppose the elevator is being accelerated upward. The path of the beam inside the elevator will no longer be a straight line—it will be curved (fig. 14.2b). To our passenger, unaware of his own acceleration, this bending of light would seem strange indeed. Einstein's answer to this dilemma was a masterpiece of logical deduction. Since the passenger has every right to believe he is standing motionlessly in Earth's gravitational field, the conclusion is inescapable: *a ray of light in a gravitational field is bent from its straight line course: it is curved.* This, in a nutshell, is the essence of general relativity.

Now it is one thing to offer this masterful explanation, but quite another to give it a precise mathematical formulation, one that allows an actual computation of the amount of bending. It is here that Einstein needed tensor analysis. As he saw it, the presence of matter at any given location warps the geometry of spacetime in the surrounding neighborhood. This warping can be expressed as a metric, the same metric that Riemann had conceived fifty years earlier, but now applied to four-dimensional spacetime. This metric has the form

$$ds^2 = \Sigma_{ij}\, g_{ij}\, dx_i\, dx_j,$$

where the $4 \times 4 = 16$ coefficients g_{ij} are expressions of the mass distribution in spacetime; they are known as the components of the symmetric gravitational tensor.[8] This metric became the mathematical centerpiece of general relativity. To quote historian of mathematics Julian Lowell Coolidge (1873–1954), "In the twentieth century reverence for Euclid has been replaced by reverence for the differential equation $ds^2 = \Sigma_{ij}\, g_{ij}\, dx_i\, dx_j$."[9]

❖   ❖   ❖

With the metric $ds^2 = \Sigma_{ij} g_{ij} dx_i dx_j$ we have come full circle. Sometime around 570 BCE, Pythagoras of Samos proved a theorem about right triangles that made his name immortal. He also reflected about the cosmos and tried to relate its workings to the laws of musical harmony. Twenty-five centuries later, another great mind, Albert Einstein, used Pythagoras's theorem, now vastly expanded and thrust into four-dimensional spacetime, to formulate his own picture of the cosmos. Had they been allowed to meet, the two sages no doubt would have found a common bond in their admiration for the beauty and harmony of the universe.

Notes and Sources

Note: "Space and Time," quoted in the epigraph, was the title of a lecture Minkowski gave before the Union of German Scientists and Physicians, meeting in Cologne on September 2, 1908. Quoted in Ronald W. Clark, *Einstein: The Life and Times* (New York: Avon Books, 1972), p. 160.

1. The word "time" here is somewhat ambiguous and really means the *rate of time*. To illustrate, two persons living in different time zones on Earth have their watches set to different times, but the watches tick at the same rate, because they travel through space at the same speed (Earth's motion around the Sun) and so belong to the same frame of reference.

2. Perhaps no one did better in explaining this derivation than Einstein himself in his book *Relativity: The Special and General Theory* (trans. Robert W. Lawson; New York: Henry Holt and Company, 1920, with numerous reprints), Appendix 1.

3. Here m is the mass of a particle *in motion*, and is itself a function of the velocity v: $m = \dfrac{m_0}{\sqrt{1 - v^2/c^2}}$, where m_0 is the particle's *rest mass*.

4. Clark, *Einstein*, p. 159.

5. Ibid., p. 160.

6. Newton himself was not happy with the concept but reluctantly adopted it, not having a better working hypothesis.

7. The aftermath of this historic event has been described many times; see, for example, Clark, *Einstein*, pp. 263–264 and 284–291. However, doubts have recently been cast on the validity of the eclipse results; see John Waller, *Einstein's Luck: The Truth Behind some of the Greatest Scientific Discoveries* (Oxford: Oxford University Press, 2002), chap. 3.

8. Actually there are only ten independent coefficients, because $g_{ij} = g_{ji}$ (see p. 173).

9. *A History of Geometrical Methods* (1940; rpt. New York: Dover, 1963), p. 78. I changed Coolidge's a_{ij} to g_{ij} to be consistent with the notation used in this chapter.

Sidebar 10

Four Pythagorean Brainteasers

If there were no puzzles to solve, there would be
no questions to ask; and if there were no questions
to be asked, what a world it would be!
 —Henry Dudeney, *Amusements in Mathematics*, p. v

Henry Ernest Dudeney (1857–1930) is regarded as England's greatest creator of mathematical puzzles. He never had formal training in mathematics, yet he developed a gift for solving difficult problems that often defied conventional thinking. He wrote six books on recreational mathematics, of which the first, *The Canterbury Puzzles*,[1] is based on characters from Chaucer's *The Canterbury Tales*. In *The Puzzles* we find the following brainteaser, quoted here verbatim:

75. - The Spider and the Fly.

Inside a rectangular room, measuring 30 feet in length and 12 feet in width and height, a spider is at a point on the middle of one of the end walls, 1 foot from the ceiling, as at *A*; and a fly is on the opposite wall, 1 foot from the floor in the centre, as shown at *B* [see fig. S10.1]. What is the shortest distance that the spider must crawl in order to reach the fly, which remains stationary? Of course, the spider never drops or uses its web, but crawls fairly.[2]

Most of us would probably suggest the "obvious," direct-line route: go 11 feet straight down from the spider's initial position to the floor, then 30 feet along the centerline of the floor, and then one foot up again to the fly. This makes a total distance of 42 feet. Yet there is a shorter route. Can you find it? So as not to spoil the fun, I will defer revealing it to Appendix H.

❖   ❖   ❖

A brainteaser known as the "Pythagorean Square" puzzle asks to combine a small square with four unequal pieces of a larger square so as

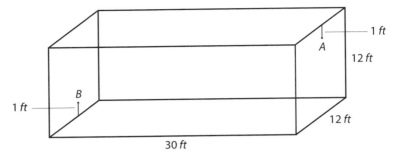

Figure S10.1. The spider and the fly

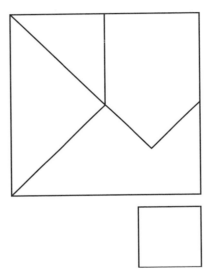

Figure S10.2. The Pythagorean Square puzzle

to form a still larger square (fig. S10.2).[3] This puzzle is indirectly related to the Pythagorean theorem insofar as the sum of the areas of two given squares equals the area of the new square. The solution—not easily found in spite of the puzzle's seeming simplicity—is given in Appendix H. Interestingly, in 1917 Dudeney suggested a similar configuration for a proof of the Pythagorean theorem (fig. S10.3).[4]

❖   ❖   ❖

The following is a relatively easy problem, taken from the ninth chapter of the *Chiu Chang Suan Shu* (Nine chapters on the mathematical art), a Chinese work dating to the Han period (206 BCE–220 CE; see p. 64):[5]

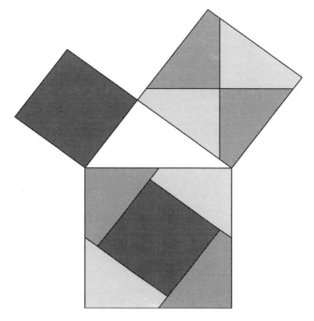

Figure S10.3. Dudeney's proof of the Pythagorean theorem

GIVEN: A tree of height 20 ch'ih has a circumference of 3 ch'ih. There is an arrow-root vine which winds seven times around the tree and reaches to the top. What is the length of the vine?

ANSWER: 29 ch'ih.

The worked-out answer can be found in Appendix H.

❖   ❖   ❖

As an encore, here is a problem from the *Lilavati* ("the beautiful"), a work written about 1150 CE by the Hindu writer Bhaskara for his daughter, reportedly to console her because the astrologers decreed she should not get married:

A snake's hole is at the foot of a pillar 15 cubits high, and a peacock is perched on its summit. Seeing a snake, at a distance of thrice the pillar's height, gliding toward his hole, he pounces obliquely upon him. Say quickly at how many cubits from the snake's hole do they meet, both proceeding an equal distance?[6]

The phrase "Say quickly" probably meant that the reader should figure out the answer mentally, without paper and pen. Can you do it? The answer is found in Appendix H.

Notes and Sources

1. (1907; rpt. New York: Dover, 1958).

2. Ibid., pp. 121–122; the solution is on pp. 221–222. A brief biographical sketch of Dudeney can be found in David Darling, *The Universal Book of Mathematics* (Hoboken, N.J.: John Wiley, 2004), p. 98. See also the Web site http://www-groups.dcs.st-and.ac.uk/~history/Mathematicians/Dudeney.html.

3. Darling, *Universal Book*, pp. 261–262 and 371. The inventor of this puzzle is not mentioned.

4. Eves, pp. 95–96.

5. Source: Frank J. Swetz and T. I. Kao, *Was Pythagoras Chinese? An Examination of Right Triangle Theory in Ancient China* (University Park, Penn.: Pennsylvania State University Press, and Reston, Va.: National Council of Teachers of Mathematics, 1977), p. 29.

6. Eves, p. 241. For more on Bhaskara and the *Lilavati*, see Smith, vol. 1, pp. 275–282.

But Is It Universal?

The Pythagorean Theorem, of course, is the
foundation of all architecture; every structure
built on this planet is based on it.
 —Michio Kaku, *Hyperspace*, p. 37

In the movie *Contact*, based on Carl Sagan's novel, a team of astronomers is
listening to faint radio signals originating from a faraway galaxy and collected
on Earth by a huge dish antenna. Suddenly they discern a pattern of dashes: --,
---, -----, ,-------, -----------, . . . : 2, 3, 5, 7, 11, . . . , the prime numbers in order!
This, the astronomers conclude, must be the calling card of a faraway civilization
trying to catch our attention. The message, the logic goes, is most likely based on
primes, those immutable building blocks of which all numbers are made. Like
the elements of the periodic table, the primes keep their identity regardless of the
circumstances under which they appear; for example, the primality of a number
does not depend on the base in which we choose to write the number. Surely,
then, these primes must have a universal existence independent of the human
mind, making them an ideal means with which to start an intergalactic dialogue.
Flush with excitement, the astronomers contact the National Security Agency
with the sensational news. The reaction? "What's a prime number?"

The primes derive their significance from the fact that every positive integer greater than 1 can be written in a unique way as a product of primes. Take
12, for example. We have $12 = 3 \times 4$; but $4 = 2 \times 2$, so $12 = 3 \times 2 \times 2$. Had we
started differently, we would still get the same prime factors: $12 = 2 \times 6$, but
$6 = 2 \times 3$, so $12 = 2 \times 2 \times 3$; except for their order, we end up with the same
primes. This fact is known as the *fundamental theorem of arithmetic*; it is at
the core of number theory, the study of integers.

There has always been a certain aura of mystery about the primes, perhaps
because many questions about them are still unsolved. In Proposition 20 of
Book IX of the *Elements*, Euclid proved that the number of primes is infinite.
But what about *twin primes*, pairs of primes of the form $(p, p + 2)$? They
seem to be quite common: (3, 5), (5, 7), (11, 13), (17, 19), . . . , (101,
103), ... One finds them even among fairly large numbers, for example

(29,879, 29,881). As of this writing, the largest known twin primes are $100,314,512,544,015 \times 2^{171,960} \pm 1$, discovered in 2006; each has 51,780 digits.[1] How many of these twin primes exist? No one knows. Most mathematicians believe there are infinitely many, but this has yet to be proved. Another famous unsolved problem is *Goldbach's conjecture*. In 1742, in a letter to Euler, Christian Goldbach (1690–1764), a mathematics professor at the Russian Imperial Academy and later a diplomat in the service of the czar's government, conjectured that every even number greater than 2 is the sum of two primes: $4 = 2 + 2$, $6 = 3 + 3$, $8 = 3 + 5$, $10 = 5 + 5$ (and also $3 + 7$), $12 = 5 + 7$, and so on. Despite computer searches that confirmed the conjecture for a huge number of even numbers, a proof is still wanting.

Until recently, primes were the exclusive domain of number theory, a branch of mathematics known for its austere beauty but seemingly lacking practical applications. But this is no longer the case. In the 1980s the primes descended from their lofty position down to the realm of earthly affairs: they now play a central role in encoding computer communication to ensure the security of the transmitted text. Moreover, with the advent of home computers, the primes reached the level of popular science. In an international project, the Great Internet Mersenne Prime Search (GIMPS), some 120,000 amateur and professional mathematicians worldwide are searching for particular primes of the form $2^n - 1$, where n itself is a prime. These are known as *Mersenne primes*, after the seventeenth-century French friar Marin Mersenne (see p. 30). It is not known how many of them exist; at the time of writing, the largest known is the gargantuan prime $2^{30,402,457} - 1$, a 9,152,052 digit number that if printed would fill some 4,000 printed pages.[2] Primes have definitely left the ivory tower of rarefied mathematics and entered the realm of everyday life.

It is easy to be carried away with prime mania, once you've been caught in their irresistible lure. But are primes really as universal as obsessed number aficionados like to think? In my four years of study as a physics undergraduate, the primes, as far as I can recall, never came up. The natural sciences, after all, are based on measurement, and when you measure a quantity, be it the length of a table, the weight of an atom, or the temperature of the Sun, you don't care a hoot if the result is a prime number or not. Primes have never played a role in the great theories that shaped modern science—Newton's universal law of gravitation, Maxwell's theory of electromagnetism, and Einstein's theory of relativity. Even Newton's great *mathematical* achievement, the invention of calculus, was influenced by his physical intuition: the calculus deals with continuously changing quantities ("fluents," as he called them), and in a world where everything is changing, there is little role for the discrete, immutable primes.

But let's take an example from the nonscientific world. Undoubtedly, the Romans were the most technologically advanced society of the ancient world. Their war machines spread fear among their enemies, and their engineering accomplishments, including buildings and bridges that still stand today, were legendary. The Romans certainly knew a great deal of practical mathematics, or else they could not have reached the level of sophistication they had attained. But pure mathematics was not their cup of tea. Though the Romans saw themselves as keepers of the Greek tradition of eminence in art and science, their contribution to mathematics was very meager. Their scientists were in the business of building physical structures, not abstract mathematical creations like primes.

So if the primes are not as universal as we often like to think of them, is there anything that is? Perhaps we should turn to geometry. In Jules Verne's classic work of science fiction, *From the Earth to the Moon* (1865), a group of space enthusiasts calling themselves the Gun Club are planning a trip to the Moon in a projectile to be fired from an enormous cannon stationed in Florida. Addressing the group, their president says:

> So much for those expeditions which I consider purely literary, since they provide no serious means for establishing relations with the luminary of the night. But I should add that some practical minds have tried to enter into serious communication with the Moon. Several years ago, for example, a German geometrician suggested that a team of scientists be sent to the steppes of Siberia. There, on the vast plains, they would set up enormous geometric figures, outlined in luminous materials, among others the square of the hypotenuse (vulgarly called "the ass's bridge" by the French). Every intelligent being, this geometrician maintained, would comprehend the scientific meaning of that figure. The Selenites [inhabitants of the Moon], if there are any, will indicate they have understood by responding with a similar figure. Once communication is established, it should be easy to create an alphabet that will make it possible to converse with the inhabitants of the Moon.[3]

The "German geometrician" has been identified by various sources as none other than Carl Friedrich Gauss.[4] It is strange indeed to think that the great Gauss would come up with such a naive idea. But in the nineteenth century many people, including well-known scientists, believed that the planets, and even our Moon, are inhabited by intelligent beings. The only question was, are these extraterrestrials advanced enough be able to communicate with us? If so, they surely must possess the same basic mathematical knowledge as we do, and by all likelihood this would include the Pythagorean theorem.

Whether true or not, the story about Gauss goes to show the unfailing confidence we humans have in mathematics. But even here we may ask some questions. Certainly the Pythagorean theorem is of fundamental importance in every branch of applied science, but let us not forget its limitations: the theorem is valid only in the plane, or on surfaces that can be developed onto a plane without distortion. On a sphere, as we saw in chapter 12, the theorem does not hold.

When the Greeks developed their geometry, they did not do so in a vacuum. Although we think of Euclid's *Elements* as a model of rigorous mathematical exposition, based entirely on logical deduction, we should bear in mind that the basic geometric concepts discussed there were all anchored in the physical world we live in. A point was an idealized dot created by a pencil on paper, a straight line represented the shortest distance between two points, and a right angle was the angle between a plumb line and the ground. Although the Greeks knew that the Earth is spherical, they created their geometry from the raw material of ordinary, day-to-day objects, and in daily life we are rarely aware of the fact that we live on a sphere. It is no wonder, then, that Euclidean geometry is by and large the geometry of flat space. Even the three-dimensional objects discussed in the *Elements*, such as the cube or the octahedron, have flat surfaces that can be constructed from two-dimensional cutouts.[5]

But let us imagine that our Earth was much smaller than it actually is, say 10 kilometers in diameter. On such a small planet, we would feel the curvature of the ground under our feet at every step we would be taking. Whether life could evolve on such a miniature world is, of course, another matter. With its exceedingly low gravity, any atmosphere surrounding such a planet would have quickly escaped into space, and without an atmosphere, life as we know it would be impossible. However, recent advances in astronomy have led to the discovery of a host of bizarre objects whose existence would have been unimaginable just a few years earlier, including well over a hundred massive planets orbiting their own stars. It is not inconceivable that a planet of just a few kilometers in diameter but having Earth's mass would hold an atmosphere and perhaps even harbor life. If advanced beings inhabit such a planet, they may have created their own geometry, tailored for their particular needs; and this geometry may well leave out the Pythagorean theorem, since it plays no significant role in their world.

But we don't have to go to other worlds to realize that even basic mathematical concepts are influenced by the physical world we live in. According to an oft-heard story, the idea of rectangular coordinates came to Descartes while lying in bed late one morning and watching a fly on the ceiling of his room. He realized that the position of the fly at any instant can be given by two numbers, the distances of the fly from each wall. But suppose Descartes had lived in an igloo. Would he still have come up with the same idea? In a dome-shaped dwelling, it is much more likely he would have discovered spherical coordinates, the system we use to locate a point on the Earth in terms of longitude and latitude. In that case, the familiar distance formula of analytic geometry would have to be replaced by a much more complicated formula. So the way we think of mathematics is very much influenced by the physical environment in which we live.

These ruminations, of course, are purely hypothetical "thought experiments" à la Einstein. Ultimately, they come down to this question: *Is mathematics a creation of the mind, or does it exist independently of it?* Or, phrased slightly differently, is it merely a tool with which we describe the physical

Figure 15.1. The Arecibo radio message (decoded)

world, or is it an inevitable *consequence* of this world? These are indeed profound questions, dwelling on the borderline between mathematics and metaphysics. They have been debated by mathematicians and philosophers for centuries, but no clear consensus has yet emerged.

❖   ❖   ❖

In a natural bowl formed by the surrounding mountains south of Arecibo, Puerto Rico, resides the world's largest radio telescope, a huge dish antenna 1,000 feet across. Operated by Cornell University since 1963, it was extensively upgraded ten years later and reinaugurated on November 16, 1974. To commemorate the occasion, an encoded radio message was beamed to the great star cluster M-13 in the constellation of Hercules, 24,000 light-years away. The message consisted of a string of 1,697 zeros and ones, which, when decoded, would provide a sort of identity card from its senders, the human beings of planet Earth. Alas, to decode it, one would have to figure out that 1,679 is a product of exactly two primes, 73×23. This would leave anyone who happens to intercept the message with just two choices: to arrange the succession of binary digits in a two-dimensional frame of either 73 rows, each consisting of 23 bits, or in 23 rows, each of 73 bits. The second arrangement reveals a crude image of the dish antenna from which the message originated, along with some details about the identity of the species who sent it (fig. 15.1; the helix-like shape at the center represents a DNA molecule). The event was described dramatically by Frank Drake, one of the founders of SETI, the Search for Extraterrestrial Intelligence organization: "By the time the ceremony and the luncheon were over, when everyone had piled back on buses to leave the site, the message had reached the vicinity of Pluto's orbit. Already it was leaving the Solar System, not after a flight of years, as with a spacecraft, but after a journey only a few hours long."[6]

So here was a message that expected its recipients to be fairly versed in mathematics, including the factoring of integers into their prime factors. Should anyone actually intercept and decode the message and care to send us a reply, we may finally be in a position to answer the question posed above—whether mathematics is the same everywhere in the universe, existing independently of its practitioners. In that case, it will be the one truly international—nay, interstellar—language, understood by anyone who has achieved mathematical literacy. Alas, we will have to wait some 48,000 years for the return message to arrive back on Earth, perhaps a few years more if the recipients have a hard time decoding it. Stay tuned!

Notes and Sources

1. According to the Web site *The Prime Pages*, http://primes.utm.edu/index.html.
2. Ibid.
3. Walter James Miller, *The Annotated Jules Verne: From the Earth to the Moon* (New York: Gramercy Books, 1995), p. 13.
4. The claim (uncorroborated as far as I know) that Gauss had made such a suggestion has surfaced from time to time—and in various variations—in a number of books. See, for example, David Darling, *The Extraterrestrial Encyclopedia: An Alphabetical Reference to All Life in the Universe* (New York: Three Rivers Press, 2000), p. 166; and Frank Drake and Dava Sobel, *Is Anyone Out There?—The Scientific Search for Extraterrestrial Intelligence* (New York: Delta, 1994), pp. 170–171. Miller, in an annotated note to *From the Earth to the Moon* (p. 13), says:

> The *German geometrician* [italics in the original] Karl F. Gauss (1777–1855), also distinguished for his work in astronomy, proposed marking the Earth with the geometric figure used to demonstrate the Pythagorean theorem: a right-angled triangle with a square on each of its three sides. The lines could be formed by wide strips of dark forest, while the spaces enclosed could be planted with some bright-colored crop like wheat. But it was another German astronomer, Joseph J. von Littrow (1781–1840), who suggested such figures be illuminated. He proposed digging ditches in the Sahara Desert in the form of, say, equilateral triangles: they could be filled with water, topped with kerosene, and lighted at night.

5. The sphere is briefly discussed in Books XII and XIII, but chiefly in relation to the five regular solids, not for the intrinsic properties of its surface.
6. Drake and Sobel, *Is Anyone Out There?*, p. 184. As to why M-13 was chosen as

the target, Drake says, "I looked at some sky charts and found that at about 1:00 P.M. on the day of the dedication, which was the time set for our ceremony, a dense cluster of some three hundred thousand stars (and possibly as many planets) would be nearly over our heads. That was our target, then: M-13, the Great Cluster in the constellation Hercules. The fact that the cluster was twenty-four thousand light-years away did not dampen my enthusiasm in the least."

16

Afterthoughts

Powers of 2 appear more frequently in
mathematics than those of any other number.
—David Wells, *The Penguin Dictionary of Curious and
Interesting Numbers*, p. 42

People often ask me: What is mathematics all about? Many think of it as the exclusive domain of numbers and computations. Those who made it through calculus know there is such a thing as "higher" mathematics, but they often wonder how it connects to the "real" world. Of course, one can always quote the famous saying, "Mathematics is what mathematicians do at night," but that will not satisfy the inquisitive mind.

So what *is* mathematics, really? I think the essence of it is the search for pattern, for structure and regularity, and for connections between seemingly unrelated objects, whether "real" or abstract. In this sense it is quite akin to art, to music in particular. Just as certain thematic and rhythmic patterns appear again and again in music, so do some algebraic expressions show up repeatedly in various areas of mathematics. Figure 16.1 shows the opening page of Mozart's Piano Concerto No. 16 in D Major, K. 451; the rhythmic motif ♩♪♫♩♩ dominates the movement from the very first bar. This motif, in its numerous variations, is a hallmark of Mozart; it can be found again and again throughout his music, including his last work, the *Requiem*, left unfinished at the time of his death. Now compare this with figure 16.2, which shows a table of integrals in a mathematical handbook; the expression $x^2 + a^2$ literally dominates the page. This same expression (perhaps with different letters) is one of the most frequently used in all of mathematics: you can find it in trigonometry, calculus, differential equations, functional analysis, and number theory—and, needless to say, in geometry, as the algebraic statement of the Pythagorean theorem.

Where do these recurring patterns come from? In Mozart's case, it probably originated in a popular four-step dance common in his time. Much of Mozart's work was commissioned for social events—balls, festive receptions, and royal dinners—where dancing took center stage. Whether consciously or

Piano Concerto No. 16 in D Major, K.451

Figure 16.1. Opening page of W. A. Mozart's Piano Concerto no. 16

INTEGRALS INVOLVING $\sqrt{x^2 + a^2}$

14.182 $\displaystyle\int \frac{dx}{\sqrt{x^2 + a^2}} = \ln(x + \sqrt{x^2 + a^2})$ or $\sinh^{-1}\frac{x}{a}$

14.183 $\displaystyle\int \frac{x\,dx}{\sqrt{x^2 + a^2}} = \sqrt{x^2 + a^2}$

14.184 $\displaystyle\int \frac{x^2\,dx}{\sqrt{x^2 + a^2}} = \frac{x\sqrt{x^2 + a^2}}{2} - \frac{a^2}{2}\ln(x + \sqrt{x^2 + a^2})$

14.185 $\displaystyle\int \frac{x^3\,dx}{\sqrt{x^2 + a^2}} = \frac{(x^2 + a^2)^{3/2}}{3} - a^2\sqrt{x^2 + a^2}$

14.186 $\displaystyle\int \frac{dx}{x\sqrt{x^2 + a^2}} = -\frac{1}{a}\ln\left(\frac{a + \sqrt{x^2 + a^2}}{x}\right)$

14.187 $\displaystyle\int \frac{dx}{x^2\sqrt{x^2 + a^2}} = -\frac{\sqrt{x^2 + a^2}}{a^2 x}$

14.188 $\displaystyle\int \frac{dx}{x^3\sqrt{x^2 + a^2}} = -\frac{\sqrt{x^2 + a^2}}{2a^2 x^2} + \frac{1}{2a^3}\ln\left(\frac{a + \sqrt{x^2 + a^2}}{x}\right)$

14.189 $\displaystyle\int \sqrt{x^2 + a^2}\,dx = \frac{x\sqrt{x^2 + a^2}}{2} + \frac{a^2}{2}\ln(x + \sqrt{x^2 + a^2})$

14.190 $\displaystyle\int x\sqrt{x^2 + a^2}\,dx = \frac{(x^2 + a^2)^{3/2}}{3}$

14.191 $\displaystyle\int x^2\sqrt{x^2 + a^2}\,dx = \frac{x(x^2 + a^2)^{3/2}}{4} - \frac{a^2 x\sqrt{x^2 + a^2}}{8} - \frac{a^4}{8}\ln(x + \sqrt{x^2 + a^2})$

14.192 $\displaystyle\int x^3\sqrt{x^2 + a^2}\,dx = \frac{(x^2 + a^2)^{5/2}}{5} - \frac{a^2(x^2 + a^2)^{3/2}}{3}$

14.193 $\displaystyle\int \frac{\sqrt{x^2 + a^2}}{x}\,dx = \sqrt{x^2 + a^2} - a\ln\left(\frac{a + \sqrt{x^2 + a^2}}{x}\right)$

14.194 $\displaystyle\int \frac{\sqrt{x^2 + a^2}}{x^2}\,dx = -\frac{\sqrt{x^2 + a^2}}{x} + \ln(x + \sqrt{x^2 + a^2})$

14.195 $\displaystyle\int \frac{\sqrt{x^2 + a^2}}{x^3}\,dx = -\frac{\sqrt{x^2 + a^2}}{2x^2} - \frac{1}{2a}\ln\left(\frac{a + \sqrt{x^2 + a^2}}{x}\right)$

14.196 $\displaystyle\int \frac{dx}{(x^2 + a^2)^{3/2}} = \frac{x}{a^2\sqrt{x^2 + a^2}}$

14.197 $\displaystyle\int \frac{x\,dx}{(x^2 + a^2)^{3/2}} = \frac{-1}{\sqrt{x^2 + a^2}}$

14.198 $\displaystyle\int \frac{x^2\,dx}{(x^2 + a^2)^{3/2}} = \frac{-x}{\sqrt{x^2 + a^2}} + \ln(x + \sqrt{x^2 + a^2})$

14.199 $\displaystyle\int \frac{x^3\,dx}{(x^2 + a^2)^{3/2}} = \sqrt{x^2 + a^2} + \frac{a^2}{\sqrt{x^2 + a^2}}$

14.200 $\displaystyle\int \frac{dx}{x(x^2 + a^2)^{3/2}} = \frac{1}{a^2\sqrt{x^2 + a^2}} - \frac{1}{a^3}\ln\left(\frac{a + \sqrt{x^2 + a^2}}{x}\right)$

14.201 $\displaystyle\int \frac{dx}{x^2(x^2 + a^2)^{3/2}} = -\frac{\sqrt{x^2 + a^2}}{a^4 x} - \frac{x}{a^4\sqrt{x^2 + a^2}}$

14.202 $\displaystyle\int \frac{dx}{x^3(x^2 + a^2)^{3/2}} = \frac{-1}{2a^2 x^2\sqrt{x^2 + a^2}} - \frac{3}{2a^4\sqrt{x^2 + a^2}} + \frac{3}{2a^5}\ln\left(\frac{a + \sqrt{x^2 + a^2}}{x}\right)$

Figure 16.2. A page from a mathematical handbook

not, the rhythmic patterns of these dances engraved themselves on his creative mind, ultimately making their way into his music.

As for the ubiquity of the expression $x^2 + a^2$ in mathematics, it can often be traced directly to the Pythagorean theorem. This is certainly true in trigonometry, where the three "Pythagorean identities" (see p. xiv) repeatedly show up. The same is also true of calculus: whenever trigonometric functions are involved, $x^2 + a^2$ is likely to show up, as the following two integrals demonstrate:

$$\int_0^\infty e^{-ax}\cos bx\, dx = \frac{a}{a^2 + b^2} \quad \text{and} \quad \int_0^\infty e^{-ax}\sin bx\, dx = \frac{b}{a^2 + b^2}$$

(where $a > 0$).

But consider now the similarly looking integrals

$$\int_0^\infty e^{-ax} J_0(bx)\,dx = \frac{1}{\sqrt{a^2 + b^2}} \quad \text{and} \quad \int_0^\infty e^{-ax} J_1(bx)\,dx = \frac{\sqrt{a^2 + b^2} - a}{b\sqrt{a^2 + b^2}}.$$

Here $J_0(x)$ and $J_1(x)$ are the *Bessel functions of order* 0 *and* 1, two "higher" functions studied in a course on differential equations.[1] These functions cannot be written in "closed" form in terms of the elementary functions (polynomials and ratios of polynomials, trigonometric and exponential functions and their inverses, and any finite combinations of them); they can only be expressed as power series in x. Bessel functions have certain outward similarities with the trigonometric functions; for example, the graphs of $J_0(x)$ and $J_1(x)$ resemble those of $\cos x$ and $\sin x$, respectively, but they are not periodic: their amplitudes diminish as x increases, and their x-intercepts are not equally spaced (fig. 16.3).[2] And yet the expression $a^2 + b^2$ mysteriously pops up in the two integrals, like a ghost of the Pythagorean theorem.

But let us return for a moment to Euclidean geometry. In principle, there is nothing that prevents us from proposing a metric different from the one actually in use, for example, a metric based on the formula $d = \sqrt[3]{x^3 + y^3}$. This metric would meet all the formal requirements we expect of distance, including the triangle inequality (see p. 160). In this metric, the "Pythagorean theorem" would take the form $a^3 + b^3 = c^3$, replacing the area of the squares built on the three sides of a right triangle with the *volume* of the corresponding cubes.[3] Nonconventional metrics are occasionally used in describing certain models of non-Euclidean geometry, but they are, by and large, artificial. The fact is that the only metric we use in practice is the quadratic Euclidean metric based on the distance formula $d = \sqrt{x^2 + y^2}$; even its generalized form $ds = \sqrt{\sum_{ij} a_{ij}\, dx_i\, dx_j}$ is still a quadratic metric.

All this makes one wonder what it is about the exponent 2 that gives it such a prominent role, not only in mathematics, but in physics as well. Newton's

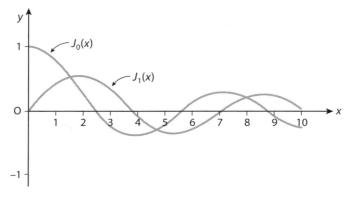

Figure 16.3. The graphs of $J_0(x)$ and $J_1(x)$

universal law of gravitation is an inverse square law, as is Coulomb's law of electricity. $E = mc^3$? It doesn't sound quite right.

But perhaps most amazing of all is the unique place of 2 in number theory, a subject as remote from the physical world as it is from trigonometry and calculus. The largest primes known today are Mersenne primes $2^n - 1$, with their associated perfect numbers $2^{n-1}(2^n - 1)$ (see p. 30). We also recall the Fermat primes $2^{2^n} + 1$ and their connection to regular polygons (p. 155).This comes at least in part from the fact that 2 is at once the smallest prime and the only even prime, giving it a special status even among the exalted primes. But no such simple explanation will do in the case of Fermat's Last Theorem. Why is it that the equation $x^n + y^n = z^n$ has integer solutions for $n = 1$ and 2, but for none other? Could 2 perhaps be the ultimate mathematical constant, exceeding even π and e in importance? I'll let the reader decide.

Notes and Sources

1. $J_0(x)$ is a solution of the differential equation $x^2 y'' + xy' + x^2 y = 0$, and $J_1(x)$ is a solution of the equation $x^2 y'' + xy' + (x^2 - 1)y = 0$. These equations are named after the German astronomer Friedrich Wilhelm Bessel (1784–1846), who introduced them in connection with his work on planetary orbits. They often appear in applications in mathematical physics, for example in the vibrations of a drumhead.

2. This has certain repercussions in music. The vibrations of a circular membrane are governed by $J_0(x)$, and the aperiodic nature of this function causes the vibrations to have nonharmonic overtones.

3. A close call is provided by the equation $6^3 + 8^3 = 9^3 - 1$.

Samos, 2005

The Pythagoreans at Samos think one should
discuss questions about goodness, justice
and expediency . . . because Pythagoras made
all those subjects his business.
  —Iambilicus of Apamea (ca. 250–330 BCE),
    Syrian author of *Life of Pythagoras*

In February 2005, my wife and I went to the island of Samos in Greece to
pay tribute to Pythagoras. Of course, we didn't expect to find a house with a
plaque, "Pythagoras was born here in the year 580 BCE"; still, I wanted to visit
the place where the young sage spent his early years and to see the sights that
shaped his vision of the universe.

The timing of our trip could hardly have been less fortunate: it was the middle of winter, the dollar was falling daily against the euro, and the heightened
concern for security made air travel an ever less enjoyable experience. But that
was the only time we could afford to go, so we called our travel agent to make
the arrangements. His first reaction was, "Why are you going in February? No
one goes to the Greek islands in midwinter." Ignoring his advice, we made it
to Athens and boarded the twin-engine, propeller-driven airplane for the fifty-
minute flight to Samos. The moment we were over the blue waters of the
Mediterranean, we could see the waves—white curls of foam whipping up the
sea—a sure sign of an impending storm. And so it was: as we approached
the island, skirting the mountain slopes that drop steeply into the sea, the pilot
struggled to keep the aircraft steady in the stiff crosswind.

The first thing we noticed after deplaning was the big sign on the airport terminal: "Aristarhos of Samos" (see plate 4). This is perhaps the only airport in the
world named after a mathematician; not Pythagoras, but his fellow Samian,
Aristarchus (ca. 310–230 BCE).[1] Aristarchus is considered the first true astronomer in history. He attempted to find the Earth's distance from the Moon
and Sun (his method was correct, but his results fell far short of the actual figures).[2] And in an extraordinary foresight, he claimed that the Sun, not the
Earth, is at the center of the universe, an idea that was far ahead of its time.

A fifteen-minute trip by taxi took us to the town of Samos, the unofficial capital of the island. By the time we settled in, the winds had picked up, and soon the whole island was in the grip of the worst winter storm in years. All flights were canceled, and the daily ferryboat from Piraeus, Greece's main port, was not allowed to leave the jetty to which it was moored, right in front of our hotel. In a small place like this, the storm was the chief news of the day. Everyone was quick to let us know that the day *before* our arrival, the weather had been picture-perfect, and it was expected to be perfect again the day *after* our departure. This, of course, did little to lift our spirits.

But we were determined not to let the 50 mph wind gusts and squalls of rain spoil our plans, so we set out to explore the town. We soon found that Pythagoras is a household name here: the main square is named after him, as is a street, a high school, and at least one hotel (see plates 5–7). And there were the usual tourist souvenirs—T-shirts, coffee mugs, and plaster busts—all showing his bearded face and the familiar drawing of his theorem. A tour booklet of Samos devoted two full pages to him, complete with a description of his theorem:

> Pythagorean Theorem: In an isosceles [*sic*] triangle the sum of the squares of the two sides is equal to the square of the hypoteneuse [*sic*].[3]

(I had to go over it twice to make sure I read it correctly, but that's what it said.) What the thousands of tourists who come here each summer make of this we couldn't tell, as the island was practically deserted; and when the crowds start arriving again in June, it is not Pythagoras that will lure them here, but the many small beaches for which the island is famous. We were odd visitors during an odd time of the year.

The next morning we took a bus to the picturesque town of Pythagorio, on the southeast side of the island. Using public transportation lets you mix with the local population, listen to their conversations (even if in a foreign language), and share their daily routines, something a rented car can never do. Sure enough, not ten minutes out of town we were treated to a local excitement of sorts. Approaching a police checkpoint, the bus was stopped and pulled aside. There was a lengthy argument between our driver and the policeman outside, who then boarded the bus with a stern look on his face. Assuming it was a security checkup, we showed him our bags, but he was uninterested, barking something in Greek which was, well, Greek to us. When we said, "English, please," he blurted "seat belts!" It was a seat-belt-enforcement inspection! The entire bus erupted in a mighty laughter, which we heartily joined. Instantly we bonded with our fellow passengers, becoming part of the crowd.

This is not a place for the faint-hearted. Everyone here was smoking, including the driver, right under the no-smoking sign at the front of the bus. Another sign, "Do not talk with the driver while the bus is in motion," did not prevent our driver from having a lively chat with his conductor. Why the driver himself couldn't collect the fare from the ten or so passengers on board I couldn't tell; perhaps it was so that two people instead of one would have employment.

Pythagorio (also spelled Pythagorion or Pythagoreio) is Samos's third-largest town. The twenty-minute trip brought us over a mountain ridge from which the coast of Turkey could be seen in the distance, shrouded in clouds. Only about 4,000 feet sea separate the two countries at this point. According to our guidebook, at one time the island was actually part of Asia Minor, but a series of earthquakes severed it from the mainland. The narrow strait marks the historic border between Europe and Asia. It was here that young Pythagoras likely crossed the sea to the town of Miletus in Asia Minor to study under the great master, Thales.

Modern Pythagorio got its name only recently; before 1955 it was called Tigani ("frying pan," so named after the shape of the jetty at the harbor). When Polycrates came to power in 550 BCE, he made it the island's capital, and its population grew to 300,000. Today it is a small town of several thousand, picturesquely tucked between the blue sea and the olive-green mountains to its north and west. To honor its famous namesake, the town erected a statue on the jetty overlooking the harbor; Pythagoras is holding a triangle in one hand, while the other is pointing up to the tip of a slanted beam, as if striving to reach it (plate 8). With a little imagination one can make up the outline of a right triangle, but nowhere was there any mention of the theorem that made Pythagoras's name immortal. I suppose few of the tourists who flock to this place in the summer really care, but I must admit I was a bit disappointed.

Pythagorio's real claim of fame, however, is the tunnel dug through the mountain by Polycrates in 524 BCE, during Pythagoras's lifetime, to safeguard the town's water supply in times of war. Named the Eupalinus Tunnel after the engineer who designed it, it is 3,400 feet long and about 5 feet high and wide, and it remained in use for over a thousand years. Its construction, as told by the historian Herodotus, was a remarkable engineering feat, the two teams digging from both ends and meeting at the center with an error of just 30 feet horizontally and 40 feet vertically.[4] When the tunnel is open to the public, one can walk along some 2,000 feet inside it, but to our great disappointment it was closed for the winter. We'll have to come again.

❖   ❖   ❖

As with all things Greek, Samos has a long and rich past. Its early history is shrouded in myth; the oldest records indicate that the Phoenicians established a permanent home there around 3000 BCE (the name Samos is believed to come from the Phoenician *Sama*, a high place). The first Ionians settled the island around 1000 BCE. In 670 BCE Samos became a democracy, and it rapidly rose to prominence. It was famous for its wines, olive groves, and, above all, its shipbuilding. Its long, swift boat, the *samaina*, was known for its exceptional speed and its ability to sail the Mediterranean down to Egypt. In 650 BCE the Samian navigator Kolaios became the first known person to sail through the Pillars of Hercules (the Strait of Gibraltar) and thus pass the boundary of the ancient world.

In 550 BCE Polycrates, who was a native of Samos, came to power and soon became the most feared Greek ruler at the time, thanks chiefly to his fleet of some 150 *samainae*. He was a great patron of art and architecture; besides the tunnel, he supervised the construction of a large harbor at the site of modern Pythagorio, and he built the great Temple of Heraion, one of the largest shrines of the ancient world, to honor the venerable goddess Hera. But his lust for power made him many enemies, and he met his end by being crucified.

In the following centuries, the island constantly shifted alliances as the battle lines between the Persians, Athenians, and Spartans moved back and forth. In 479 BCE the Greeks defeated the Persian navy at the Strait of Mykale, ending Persian domination for good. In 129 BCE Samos became part of the Roman Empire. Emperor Augustus often visited the island in winter, and as the historian Plutarch tells us, it was the favorite retreat of Anthony and Cleopatra.

After the rise of Christianity, Samos was ruled by the Venetians and Genoese; the latter handed it to the Turks in 1453. For the next century the island was deserted, its inhabitants fleeing to the nearby island of Chios. Then the Ottomans repopulated it with their own. In response to this domination, the Samians rose in revolt, and in 1830 they defeated the Turks in the second battle of Mykale. Although the Great Powers excluded Samos from Greece, it was granted semiautonomous status under the "hegemony of the prince of Samos," a Christian governor appointed by the Turkish Sultan. In 1913, with the Ottoman Empire in decline following its defeat in the Balkan Wars, Samos was reunited with Greece.[5]

❖   ❖   ❖

On our last day we took the bus to Karlovasi, the largest town of Samos and the seat of its only university. The winds were still blowing with ferocious force, so we limited ourselves to a brief tour of the town. At the central square we found the statue of Aristarchus, the island's other scientist of fame (plate 9). The inscription on the marble pedestal, in English and Greek, had this to say:

Aristarchos of Samos, 320–250 BC. First to Discover the Earth revoltes [*sic*] around the Sun. Copernicus copied Aristarchos 1530 AD.

Aristarchus has often been called the Copernicus of Antiquity, but whether Copernicus "copied" from him should perhaps be left for historians to decide.

❖   ❖   ❖

It was time to say *adieu* to the island. As our airplane took off, it skirted Mount Kerkis, 4,711 feet high, the tallest peak on the island. I recalled the words of our guidebook:

In the pitch black nights of winter, when the fishermen pass by the wind-buffeted and sheer slopes of Mt. Kerkis, they say they see a light up on the peak, which like a lighthouse guides them on a safe course during a storm. They say that the light is the spirit of Pythagoras. Pythagoras was born on Samos nearly 2,500 years ago and benefited the world with his philosophy and mathematics. He still lives in the hearts and souls of the fishermen of Samos.[6]

Notes and Sources

1. I have used the common English spelling Aristarchus rather than Aristarchos, although the latter may be closer to the Greek pronunciation.

2. See *Trigonometric Delights*, pp. 63–65.

3. *Samos: The Island of Pythagoras* (Koropi, Greece: Michael Toubis Publications S.A., 1995), p. 17.

4. Some interesting mathematics was involved in this feat; see Bartel L. van der Waerden, *Science Awakening: Egyptian, Babylonian and Greek Mathematics* (1954; trans. Arnold Dresden, 1961; rpt. New York: John Wiley, 1963), pp. 102–104.

5. This brief historical sketch is based on Dana Facaros, *Greek Islands* (London: Cadogan Books, 1998), pp. 517–518.

6. *Samos*, p. 30.

How did the Babylonians Approximate $\sqrt{2}$?

How did the Babylonians find the remarkably close approximation of $\sqrt{2}$ in YBC 7289? It would be nice if we had a clay tablet with exact instructions for finding square roots, but such is not the case. It is, however, reasonably safe to assume that they used the iterative formula

$$x_{i+1} = \frac{1}{2}(x_i + \frac{2}{x_i}), \quad i = 0, 1, 2, 3, \ldots,$$

sometimes known as the *Newton-Raphson formula*. Start with an initial guess $x_0 > 0$ and put it in the formula, producing the value $x_1 = (x_0 + 2/x_0)/2$. Put this new value back into the formula to produce $x_2 = (x_1 + 2/x_1)/2$, and so on. The values of x_i so obtained will all be in excess of $\sqrt{2}$ (see below); but as i increases, they rapidly converge to $\sqrt{2}$. For example, choosing $x_0 = 1.5$, we get $x_1 = 1.4166667$, $x_2 = 1.4142157$, $x_3 = 1.4142136$, and so on. After only three steps, we reached a value correct to six places. Indeed, it is this value, engraved in sexagesimal digits as 1;24,51,10, that appears on YBC 7289 (see p. 6).

The same procedure can be used to find the square root of any positive number a; the formula is $x_{i+1} = (x_i + a/x_i)/2, i = 0, 1, 2, 3, \ldots$. To justify it, we note that if the initial guess x_0 was greater than $\sqrt{a}$, then a/x_0 is smaller than $\sqrt{a}$, and vice versa. In either case, taking their average $(x_0 + a/x_0)/2$ produces a better approximation of the desired value. A well-known theorem says that the arithmetic mean of two positive numbers is never less than their geometric mean; that is, $(a + b)/2 \geq \sqrt{ab}$, with equality holding if and only if $a = b$. Applying this inequality to the expression above, we get

$$\frac{1}{2}(x_i + \frac{a}{x_i}) \geq \sqrt{x_i \cdot \frac{a}{x_i}} = \sqrt{a},$$

showing that from x_1 on, all approximations are in excess of $\sqrt{a}$ (unless we happened to start with $x_0 = \sqrt{a}$, in which case all x_i will be equal to $\sqrt{a}$).

To show that $\frac{1}{2}(x_i + a/x_i)$ tends to $\sqrt{a}$ as $i \to \infty$, we need to show that the error $\varepsilon_i = x_i - \sqrt{a}$ of the ith approximation tends to zero as $i \to \infty$. Replacing for a moment i by $i + 1$, we have

$$\varepsilon_{i+1} = x_{i+1} - \sqrt{a} = \frac{1}{2}(x_i + a / x_i) - \sqrt{a}.$$

But we showed above that $x_i > \sqrt{a}$ for all $i = 1, 2, \ldots$, and so $a/x_i < a/\sqrt{a} = \sqrt{a}$. Putting this back in the last equation, we find that

$$\varepsilon_{i+1} < \frac{1}{2}(x_i + \sqrt{a}) - \sqrt{a} = \frac{1}{2}(x_i - \sqrt{a}) = \frac{\varepsilon_i}{2}.$$

This shows that with each successive approximation, the error decreases to less than one-half its previous value. But this means that as $i \to \infty$, $\varepsilon_i \to 0$, and consequently $x_i \to \sqrt{a}$.

The assumption that this "divide and average" procedure was indeed the Babylonian method of approximating square roots is bolstered by the existence of clay tablets listing the reciprocals of numbers, making it relatively easy for a scribe to do a division problem by turning it into multiplication.[1]

Notes and Sources

1. See Otto Neugebauer, *The Exact Sciences in Antiquity* (1957; rpt. New York: Dover, 1969), pp. 32–34. Other sources mention a related formula, $\sqrt{a^2 + b} \sim a + b / 2a$ when $b \ll a$, as a possible Babylonian method of finding square roots; see Bartel L. van der Waerden, *Science Awakening: Egyptian, Babylonian and Greek Mathematics* (1954; trans. Arnold Dresden, 1961; rpt. New York: John Wiley, 1963), pp. 37 and 44–45.

Pythagorean Triples

$\mathbf{I}$n chapter 1 we showed that, given two integers u and v with $u > v$, u and v relatively prime (having no common factor other than 1) and of opposite parity (one even, the other odd), the integers

$$a = 2uv, \quad b = u^2 - v^2, \quad c = u^2 + v^2 \tag{1}$$

form a primitive Pythagorean triple (a, b, c); that is,

$$a^2 + b^2 = c^2. \tag{2}$$

We now prove the converse: for every primitive Pythagorean triple (a, b, c), there exist integers u and v, with $u > v$ and u and v relatively prime, such that equations (1) are fulfilled.

We first note that for (a, b, c) to be primitive—that is, for a, b and c to have no common factors other than 1—one of a or b must be even, the other odd, and c must be odd. First we show that a and b cannot both be even. For if they were, so would a^2 and b^2 and hence their sum, which is c^2. But then c itself would be even, and so a, b, and c would have the common factor 2, contrary to the assumption that (a, b, c) is a primitive triple.

Now we show that a and b cannot both be odd. Suppose they were. Then we can write $a = 2m + 1$, $b = 2n + 1$, for some integers m and n. Squaring and adding, we get

$$a^2 + b^2 = (2m + 1)^2 + (2n + 1)^2 = 4(m^2 + n^2 + m + n) + 2 = 4r + 2,$$

where $r = m^2 + n^2 + m + n$ is an integer. But this means that $a^2 + b^2$, and therefore c^2, leaves a remainder 2 upon dividing by 4. On the other hand, since we assumed that a and b are odd, so are a^2 and b^2. Now the sum of two odd numbers is even, so $a^2 + b^2 = c^2$ must be even. But then c itself must be even, and so $c = 2s$ for some integer s. Now the square of an even number is always divisible by 4 (indeed, $(2s)^2 = 4s^2$, which means that it leaves a remainder 0 upon dividing by 4). Since a number cannot at once leave remainders 0 and 2 upon dividing by 4, we arrived at a contradiction, proving that a and b cannot both be odd. Thus one of them must be even and the other odd. And because equation (2) is symmetric with respect to a and b, we can assume without loss of generality that a is even and b odd.

So let us write $a = 2t$ and put this into equation (2):

$$(2t)^2 + b^2 = c^2,$$

and so

$$(2t)^2 = c^2 - b^2 = (c + b)(c - b),$$

which we can rewrite as

$$t^2 = \frac{c+b}{2} \cdot \frac{c-b}{2}. \tag{3}$$

Note that since b and c are both odd, their sum and difference are both even; consequently, each of the factors in equation (3) is an integer, and these integers are relatively prime (if they had a common factor, then so would their sum $(c + b)/2 + (c - b)/2 = c$ and their difference $(c + b)/2 - (c - b)/2 = b$; and in view of equation (2), this common factor would also be shared by a, contrary to the assumption that (a, b, c) is primitive).

Now, since the left side of equation (3) is a perfect square, so is the right side; and because the two factors on the right side are relatively prime, their prime factorization must include each prime an even number of times; that is, each of the factors on the right side is itself a perfect square.

So let us write

$$\frac{c+b}{2} = u^2, \qquad \frac{c-b}{2} = v^2, \tag{4}$$

where u and v are relatively prime, and $u > v$. Adding and subtracting equations (4), we get $c = u^2 + v^2$ and $b = u^2 - v^2$; and since both b and c are odd, either of these last equations shows that u and v have opposite parities. Finally, substituting equations (4) into equation (3), we get $t^2 = u^2 v^2$, from which $t = uv$ and thus $a = 2uv$. This completes the proof.

Sums of Two Squares

Related to Pythagorean triples is the question of which integers can be written as the sum of two squares (we consider only nonnegative integers). Clearly some integers can be so written, but others cannot. For example, $5 = 1^2 + 2^2$, but 6 is not the sum of two squares (of course, every perfect square is the sum of two squares if we allow 0 to be counted). Our goal is to find a criterion that will determine whether or not a given integer can be written as a sum of two squares.

In what follows, we need to define a concept from number theory. We say that two integers a and b are *congruent modulo m* if they leave the same remainder after division by m; we write $a \equiv b \pmod{m}$. For example, $7 \equiv 11$ (mod 4), because when 7 or 11 are divided by 4, each leaves a remainder 3. Similarly, $13 \equiv 3$ (mod 5), $15 \equiv 0$ (mod 5), and so on. Simply stated, $a \pmod{m}$ tells us how far a is from an exact division by m.

In elementary number theory it is proved that the "mod" operation fulfills many of the ordinary rules of algebra; for example, if $a \equiv b \pmod{m}$ and $c \equiv d \pmod{m}$, then $a + c \equiv b + d \pmod{m}$, and similarly for multiplication. As a consequence, modulo arithmetic (sometimes called "clock arithmetic" because of the analogy to the hours of a clock, a mod 12 system) is very similar to ordinary arithmetic. For instance, squaring the congruence $13 \equiv 3$ (mod 5), we get $169 \equiv 9$ (mod 5), which it true because 169 and 9 each leaves a remainder 4 after dividing by 5. With this in mind, we now prove the following theorem:

> *A positive integer a is the sum of two squares only if a is* not *congruent to 3 (mod 4); that is, only if a does not leave a remainder 3 after dividing by 4.*

Proof:

Upon dividing by 4, an integer a can only leave a remainder 0, 1, 2, or 3; that is, $a \equiv 0, 1, 2, 3 \pmod{4}$. Squaring, we get $a^2 \equiv 0, 1, 4, 9$ (mod 4); but 4 (mod 4) $\equiv 0$ and 9 (mod 4) $\equiv 1$, so $a^2 \equiv 0, 1 \pmod{4}$. The same is true for any other integer b: $b^2 \equiv 0, 1 \pmod{4}$. Adding the two congruencies, we get $a^2 + b^2 \equiv 0, 1, 2$ (mod 4). Thus, the sum of two squares can never be congruent to 3 (mod 4).

Note that the theorem is a *necessary*, but not a *sufficient*, condition for a number to be the sum of two squares; that is, a number may be $\not\equiv 3$ (mod 4) and still not be a sum of two squares, as the example of 12 ($\equiv 0$ (mod 4)) shows. Of the first twenty positive integers, only eight are sums of two squares: $2 = 1^2 + 1^2$, $5 = 1^2 + 2^2$, $8 = 2^2 + 2^2$, $10 = 1^2 + 3^2$, $13 = 2^2 + 3^2$, $17 = 1^2 + 4^2$, $18 = 3^2 + 3^2$, and $20 = 2^2 + 4^2$ (to which we may add 1, 4, 9, and 16 if we allow 0 to count as a square).

The following theorem, which we give here without proof, provides a necessary *and* sufficient condition for a number to be the sum of two squares:

> *An integer a is the sum of two squares if and only if any prime congruent to 3 (mod 4) in the prime factorization of a appears an even number of times in the factorization.*[1]

This theorem explains why 12 is not the sum of two squares. The prime factorization of 12 is $2 \times 2 \times 3$, and since the prime $3 \equiv 3$ (mod 4) appears only once, 12 cannot be the sum of two squares. On the other hand, $18 = 2 \times 3 \times 3$, and since 3 appears twice, 18 is the sum of two squares.

The following theorem, due to Diophantus (see p. 57), makes it possible to "build up" sums of two squares from simpler ones:

> *If two integers are each the sum of two squares, so is their product.*

Proof:

Let $p = a^2 + b^2$, $q = c^2 + d^2$. Then

$$pq = (a^2 + b^2)(c^2 + d^2)$$

$$= a^2c^2 + a^2d^2 + b^2c^2 + b^2d^2$$

$$= (ac)^2 + 2(ac)(bd) + (bd)^2 + (ad)^2 - 2(ad)(bc) + (bc)^2$$

$$= (ac + bd)^2 + (ad - bc)^2, \tag{1}$$

a sum of two squares. Alternatively, we also have

$$pq = (a^2 + b^2)(c^2 + d^2) = (ac - bd)^2 + (ad + bc)^2, \tag{2}$$

raising the possibility that a number can be written as a sum of two squares in two different ways. As an example, take the numbers 5 and 10, each a sum of two squares. We have

$$50 = 5 \times 10 = (1^2 + 2^2) \times (1^2 + 3^2)$$

$$= (1 \times 1 + 2 \times 3)^2 + (1 \times 3 - 2 \times 1)^2 = 7^2 + 1^2,$$

or

$$50 = (1 \times 1 - 2 \times 3)^2 + (1 \times 3 + 2 \times 1)^2 = (-5)^2 + 5^2 = 5^2 + 5^2.$$

In fact, 50 is the smallest integer that can be written as a sum of two squares in two ways; the next such integer is $65 = 1^2 + 8^2 = 4^2 + 7^2$.

When a *perfect square* is the sum of two squares, we have a Pythagorean triple (a, b, c), with $c^2 = a^2 + b^2$. It is possible to construct such triples by factoring c into smaller factors, each of which is itself a sum of two squares. For example, to find a triple with $c = 481$, we write 481 as the sum of two squares in two ways:

$$481 = 13 \times 37 = (2^2 + 3^2) \times (1^2 + 6^2)$$
$$= (2 \times 1 + 3 \times 6)^2 + (2 \times 6 - 3 \times 1)^2 = 20^2 + 9^2,$$

and also

$$481 = (2 \times 1 - 3 \times 6)^2 + (2 \times 6 + 3 \times 1)^2 = 16^2 + 15^2.$$

Therefore,

$$481^2 = (20^2 + 9^2) \times (16^2 + 15^2)$$
$$= (20 \times 16 + 9 \times 15)^2 + (20 \times 15 - 9 \times 16)^2 = 455^2 + 156^2,$$

or again,

$$481^2 = (20 \times 16 - 9 \times 15)^2 + (20 \times 15 + 9 \times 16)^2 = 185^2 + 444^2.$$

We can, in fact, get two more triples:

$$481^2 = 13^2 \times 37^2 = (5^2 + 12^2) \times (12^2 + 35^2)$$
$$= (5 \times 12 + 12 \times 35)^2 + (5 \times 35 - 12 \times 12)^2 = 480^2 + 31^2,$$

and

$$481^2 = (5 \times 12 - 12 \times 35)^2 + (5 \times 35 + 12 \times 12)^2 = 360^2 + 319^2.$$

The last of these triples, $(360, 319, 481)$, appears on the sixth line of the Babylonian clay tablet Plimpton 322 (see p. 9), and it is quite possible that some of the other large triples in this table where obtained in the same way. The four triples can be represented as four right triangles inscribed in a circle with diameter 481 (fig. C.1).

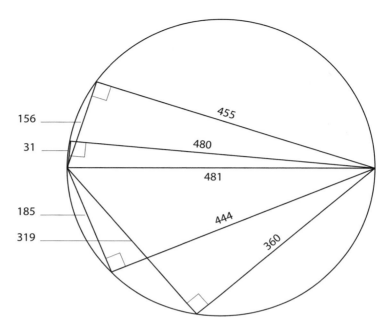

Figure C.1. The Pythagorean triples (31, 480, 481), (156, 455, 481), (185, 444, 481), and (319, 360, 481), represented geometrically

Notes and Sources

1. For a proof, see Charles Vanden Eynden, *Elementary Number Theory* (New York: McGraw-Hill, 1987), pp. 232–237.

A Proof That $\sqrt{2}$ Is Irrational

We do not know how the Pythagoreans proved the irrationality of $\sqrt{2}$, but it is likely their proof was based on geometric arguments.[1] We give here a proof based on the fundamental theorem of arithmetic.

We follow the indirect method. Assume $\sqrt{2}$ *is* rational, so it can be written as a ratio of two integers:

$$\sqrt{2} = m/n. \tag{1}$$

Squaring, we get

$$m^2 = 2n^2. \tag{2}$$

By the fundamental theorem of arithmetic, m and n can be factored uniquely into their prime factors, so let $m = p_1 p_2 \ldots p_r$ and $n = q_1 q_2 \ldots q_s$. Putting this back in equation (2), we get

$$(p_1 p_2 \ldots p_r)^2 = 2(q_1 q_2 \ldots q_s)^2,$$

or

$$p_1 p_1 p_2 p_2 \ldots p_r p_r = 2 q_1 q_1 q_2 q_2 \ldots q_s q_s. \tag{3}$$

Now among the primes p_i and q_i, the prime 2 *may* occur (it will occur if either m or n is even). If it does occur, it must appear an *even* number of times on the left side of equation (3) (since each prime there appears twice), and an *odd* number of times on the right side (because 2 already appears there once). This is true even if 2 does *not* occur among the p_i or q_i; in that case 2 will not appear at all on the left side but will occur once on the right side. In either case we have a contradiction: since the factorization into primes is unique, the prime 2 cannot appear an even number of times on one side of the equation and an odd number on the other. Thus equation (3), and therefore equation (1), cannot be true: $\sqrt{2}$ cannot be written as the ratio of two integers, and must therefore be irrational.

The same proof can be used to show that the square root of every prime number is irrational.

In Greek parlance, an irrational number is *incommensurable* with the unit— the two numbers do not have a common measure. Thus, if $\sqrt{2}$ were commensurable with 1, there would exist a line segment of length p that goes an exact number of times into both $\sqrt{2}$ and 1, say $\sqrt{2} = mp$ and $1 = np$, where m and n are positive integers. Dividing the first equation by the second, we get $\sqrt{2} = mp/np = m/n$, a rational number, contradicting the fact that $\sqrt{2}$ is irrational.

Notes and Sources

1. See Eves, p. 84; for alternative proofs, see pp. 82–83 and 356–357.

Archimedes' Formula for Circumscribing Polygons

We wish to find a formula for the length of a side s_{2n} of a $2n$-sided regular polygon circumscribing a circle in terms of the length of a side s_n of an n-sided circumscribing polygon. Let the circle have radius 1, as in figure E.1 (the figure shows a square and a regular octagon, but our proof is entirely general). Let AB be one side of the $2n$-gon, and let its midpoint be C. AB is tangent to the circle at C, so $\angle OCB = 90°$. Extend OC to the vertex E of the

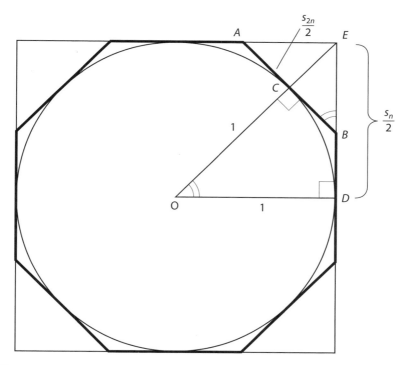

Figure E.1. The relation between s_n and s_{2n} for circumscribing regular polygons

n-gon. Since $OD \perp ED$ and $BC \perp EC$, we have $\angle EOD = \angle EBC$. Therefore, triangles EOD and EBC are similar, and so

$$\frac{ED}{OD} = \frac{EC}{BC}. \tag{1}$$

Now, $ED = s_n/2$, $OC = OD = 1$, $EC = OE - OC = \sqrt{OD^2 + ED^2} - OC = \sqrt{1^2 + (s_n/2)^2} - 1$, and $BC = s_{2n}/2$. Putting all this back in equation (1), we get

$$\frac{(s_n/2)}{1} = \frac{\sqrt{1^2 + (s_n/2)^2} - 1}{(s_{2n}/2)}. \tag{2}$$

Solving equation (2) for s_{2n} in terms of s_n, we get Archimedes's formula,

$$s_{2n} = \frac{2\sqrt{4 + s_n^2} - 4}{s_n}.$$

Proofs of Some Formulas from Chapter 7

1. To rectify the logarithmic spiral, we use the formula for arc length in polar coordinates (see p. 87):

$$s = \int_{\theta_1}^{\theta_2} \sqrt{r^2 + (dr/d\theta)^2}\, d\theta.$$

The polar equation of the spiral is $r = e^{a\theta}$, where $a =$ constant. Differentiating, we get $dr/d\theta = ae^{a\theta} = ar$. Thus,

$$s = \int_{\theta_1}^{\theta_2} \sqrt{r^2 + (ar)^2}\, d\theta = \sqrt{1 + a^2} \int_{\theta_1}^{\theta_2} e^{a\theta} d\theta$$

$$= \frac{\sqrt{1 + a^2}}{a}(e^{a\theta_2} - e^{a\theta_1}), \tag{1}$$

where θ_1 and θ_2 are the lower and upper limits of integration, respectively.

Let us assume $a > 0$; that is, r increases with θ (a left-handed spiral; see fig. F.1). Taking θ_2 in equation (1) to be fixed and letting $\theta_1 \to -\infty$, we have $e^{a\theta_1} \to 0$, and so

$$s_\infty = \frac{\sqrt{1 + a^2}}{a} e^{a\theta_2} = \frac{\sqrt{1 + a^2}}{a} r_2. \tag{2}$$

Thus, for a left-handed spiral, the arc length from any point on the spiral to the pole has a finite value. If the spiral is right-handed ($a < 0$), we let $\theta_1 \to +\infty$, arriving at a similar conclusion.

The expression on the right side of equation (2) can be interpreted geometrically. Since a can have any real value, positive or negative, we may substitute $a = \cot \alpha$ and use the identity $1 + \cot^2\alpha = \csc^2\alpha$ to rewrite the right side of equation (2) as

$$s_\infty = r \sec \alpha, \tag{3}$$

where we dropped the subscript "2" under r. Referring to figure F.2 and taking P as the point from which we measure the arc length to the pole, we have $\sec \alpha = PT/OP = PT/r$. Hence, $PT = r \sec \alpha = s_\infty$; that is, the distance along

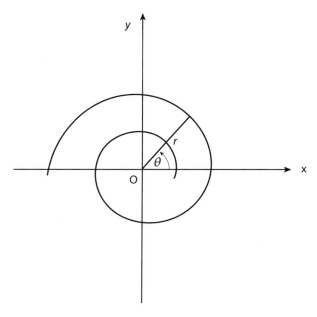

Figure F.1. A left-handed logarithmic spiral

the spiral from P to the pole is equal to the length of the tangent line to the spiral between P and T.[1]

2. To rectify the cycloid, we use its parametric equations (see p. 88):

$$x = a(\theta - \sin \theta), \; y = a(1 - \cos \theta), \tag{4}$$

where θ ranges from 0 to 2π. The length of one arch is therefore

$$s = \int_0^{2\pi} \sqrt{dx^2 + dy^2} = \int_0^{2\pi} \sqrt{(dx/d\theta)^2 + (dy/d\theta)^2} \, d\theta. \tag{5}$$

From the parametric equations we have $dx/d\theta = a(1 - \cos \theta)$, $dy/d\theta = a \sin \theta$, so

$$\sqrt{(dx/d\theta)^2 + (dy/d\theta)^2} = a\sqrt{(1 - \cos \theta)^2 + \sin^2 \theta}$$

$$= a\sqrt{2(1 - \cos \theta)} = 2a \sin \frac{\theta}{2},$$

where we used the identity $\sin^2 \frac{\theta}{2} = \frac{1 - \cos\theta}{2}$. Putting this back in equation (5), we get

$$s = 2a\int_0^{2\pi} \sin \frac{\theta}{2} \, d\theta = -4a \cos \frac{\theta}{2}\Big|_0^{2\pi} = -4a(\cos\pi - 1) = 8a. \tag{6}$$

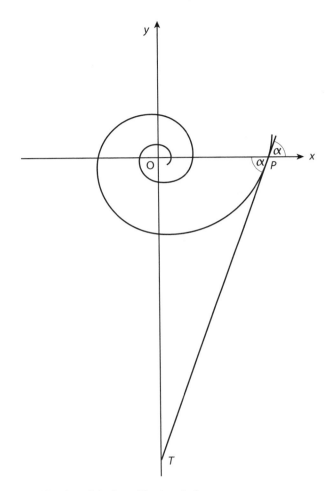

Figure F.2. Rectification of the logarithmic spiral

Thus the arc length of one arch of the cycloid is eight times the radius of the generating circle; interestingly, π does not enter the result (it does appear in the expression for area under one arch, $3\pi a^2$).

3. To rectify the astroid, we start with its implicit equation, $x^{2/3} + y^{2/3} = 1$. Differentiating implicitly, we get $\frac{2}{3}x^{-1/3} + \frac{2}{3}y^{-1/3}y' = 0$, from which $y' = (y/x)^{1/3}$. To find the arc length, we need the expression $\sqrt{1 + y'^2}$. We have

$$1 + y'^2 = 1 + (y/x)^{2/3} = (x^{2/3} + y^{2/3})/x^{2/3} = 1/x^{2/3} = x^{-2/3},$$

and so $\sqrt{1+y'^2} = x^{-1/3}$. Therefore,

$$s = \int_0^1 \sqrt{1+y'^2}\,dx = \int_0^1 x^{-1/3}dx = \frac{3}{2}x^{2/3}\Big|_0^1 = 3/2. \tag{7}$$

That is, the arc length of one-quarter of the astroid is $1\frac{1}{2}$ times the length of the generating rod.[2]

Notes and Sources

1. More on the logarithmic spiral can be found in *e: The Story of a Number*, chap. 11, pp. 135–139, and Appendix 6, from which the material above is taken.

2. For additional properties of the astroid, see *Trigonometric Delights*, pp. 98–101.

Deriving the Equation $x^{2/3} + y^{2/3} = 1$

In chapter 10 we obtained the line equation of the curve formed by a ladder of length 1 taking all possible positions as its top end slides down a wall: $1/\alpha^2 + 1/\beta^2 = 1$, whose graph is an astroid. Our goal is to find the rectangular equation of this curve.

A curve in two dimensions is given implicitly by an equation of the form $f(x, y) = 0$. A *family of curves* with a common property can be described by the equation

$$f(x, y, c) = 0, \tag{1}$$

where c is a parameter whose value changes continuously from one curve to another. When two members of the same family intersect, we have $f(x, y, c_1) = f(x, y, c_2)$, which we can write as

$$\frac{f(x, y, c_1) - f(x, y, c_2)}{c_1 - c_2} = 0.$$

This equation is true as long as $c_1 \neq c_2$; in fact, its left side has the appearance of a difference quotient, $\Delta f/\Delta c$. As $c_1 \to c_2$, the equation becomes

$$\frac{\partial f}{\partial c} = 0, \tag{2}$$

where we used the partial-derivative symbol ∂ to indicate that x and y remain fixed during the differentiation. Equations (1) and (2), when taken together, are the parametric equations of the *envelope* formed by the continuous intersection of neighboring curves of the family. By eliminating the parameter c between these equations, we can find the rectangular equation of the envelope.

In our example, the envelope is generated by the ladder as it assumes all possible positions. The equation of the ladder is

$$\alpha x + \beta y = 1, \tag{3}$$

where α and β are related through the line equation

$$1/\alpha^2 + 1/\beta^2 = 1. \tag{4}$$

Solving equation (4) for β, we get $\beta = \alpha/\sqrt{\alpha^2 - 1}$. Substituting this into equation (3) gives us

$$\alpha x + \frac{\alpha}{\sqrt{\alpha^2 - 1}} y = 1$$

or

$$\alpha \left[x + \frac{y}{(\alpha^2 - 1)^{1/2}} \right] = 1. \tag{5}$$

This is the equation of the family of tangent lines, with α serving as the parameter c. Differentiating it with respect to α and simplifying, we get

$$x - \frac{y}{(\alpha^2 - 1)^{3/2}} = 0. \tag{6}$$

We must now eliminate α between equations (5) and (6). From equation (6) we get

$$(\alpha^2 - 1)^{3/2} = y/x, \text{ so } \alpha^2 - 1 = (y/x)^{2/3}, \alpha^2 = 1 + (y/x)^{2/3} = (x^{2/3} + y^{2/3})/x^{2/3},$$

and finally $\alpha = (x^{2/3} + y^{2/3})^{1/2}/x^{1/3}$. Putting this back into equation (5), we get,

$$\frac{(x^{2/3} + y^{2/3})^{1/2}}{x^{1/3}} \left[x + \frac{y}{(y/x)^{1/3}} \right] = 1. \tag{7}$$

The expression in the brackets can be simplified to

$$x + y(y/x)^{-1/3} = x + y^{2/3}x^{1/3} = x^{1/3}(x^{2/3} + y^{2/3}),$$

so equation (7) becomes

$$\frac{(x^{2/3} + y^{2/3})^{1/2}}{x^{1/3}} \cdot x^{1/3}(x^{2/3} + y^{2/3}) = 1.$$

Canceling $x^{1/3}$ and collecting powers of $(x^{2/3} + y^{2/3})$, this equation simplifies to $(x^{2/3} + y^{2/3})^{3/2} = 1$, from which we finally get

$$x^{2/3} + y^{2/3} = 1.$$

This rather long transformation process should make it clear that, in some cases, line coordinates are better suited to describe a curve than rectangular coordinates.

Solutions to Brainteasers

The Spider-and-Fly brainteaser (p. 197) can be solved by flattening the rectangular box, just as when you flatten a used shoe box before it goes to the recycling bin. This can be done in essentially three ways (see fig. H.1, in which only the relevant sides are shown). In case (a) the distance is $d = 1 + 30 + 11 = 42$ ft. For the remaining cases, we need the Pythagorean theorem. In case (b), the horizontal distance between the spider and the fly is $1 + 30 + 6 = 37$ ft, and the vertical distance is $6 + 11 = 17$ ft, so $d = \sqrt{37^2 + 17^2} = \sqrt{1658} \sim 40.7$ ft. In case (c), the horizontal distance is $1 + 30 + 1 = 32$ ft and the vertical distance is $6 + 12 + 6 = 24$ ft, so $d = \sqrt{32^2 + 24^2} = \sqrt{1600} = 40$ ft. Thus, case (c) provides the shortest distance.

You may ask: Since the spider cannot fly directly to its prey, how would it know what route to follow? To answer this, we need a little trigonometry. Let the angle between the direction of the path and the perpendicular to the edge closest to the spider be α. From figure H.1c we have $\tan \alpha = 24/32 = 3/4 = 0.75$, so $\alpha \sim 36.9°$. Thus the spider would have to crawl *up* at an angle of about 37° with the vertical to reach the ceiling, then set on a bearing of 37° relative to the "north" (the direction of the long side of the room) and follow this bearing across the ceiling, front wall, and floor, then go up again on the opposite side wall at 37° to the vertical, which will finally bring the spider to its prey (assuming the latter had been quietly awaiting its predator all this time). The entire route takes the spider through five of the six faces of the room; it demonstrates that the "direct" route, case (a), is not always the shortest possible route—the *geodesic*—for the particular surface under consideration.[1]

The solution to the Pythagorean Square puzzle (p. 197) is shown in figure H.2.

❖   ❖   ❖

To find the length of the vine, we note that it winds exactly seven times around the tree, so that its top and bottom ends are lined up. Assuming the tree has a

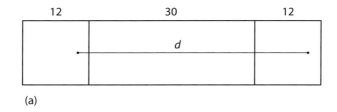

(a)

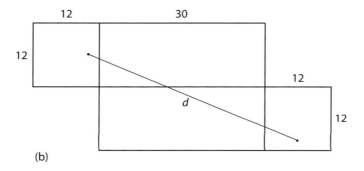

(b)

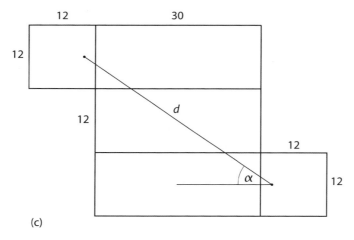

(c)

Figure H.1. The Spider and the Fly: the three possible routes

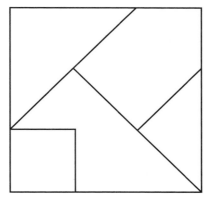

Figure H.2. Solution to the Pythagorean Square puzzle

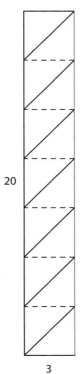

20

3   Figure H.3. The Winding Vine

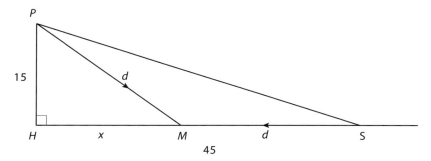

Figure H.4. The Snake and the Peacock

cylindrical trunk, we can unwrap it to get the zigzag pattern shown in figure H.3, where each pair of horizontal points represents one and the same point on the trunk. Each triangle has a base of 3 ch'ih and height of 20/7 ch'ih, so each hypotenuse has a length $\sqrt{3^2 + (20/7)^2} = \frac{\sqrt{841}}{7} = 29/7$; and since there are seven of these triangles, all congruent, the total length is 29 ch'ih.

❖ ❖ ❖

To find the distance from the snake's hole to the point where the snake and peacock meet, we refer to figure H.4. The peacock is at P, the hole at H, the snake is sighted at S, and the meeting point is at M. Let $HM = x$, $PM = MS = d$. We have $d^2 = 15^2 + x^2$ and $d + x = 45$. Rewriting the first equation as $d^2 - x^2 = (d + x)(d - x) = 45 \, (d - x) = 15^2 = 225$, we get $d - x = 5$. Solving the pair of equations $d + x = 45$ and $d - x = 5$, we get $x = 20$ cubits and $d = 25$ cubits. This shows that triangle PHM is the Pythagorean triangle $(15, 20, 25)$, that is, the triangle $(3, 4, 5)$ magnified fivefold.

Notes and Sources

1. A similar situation happens on a globe. The shortest distance between two cities with the same latitude is not along their common circle of latitude, but along an arc of the great circle connecting the two cities. For example, the great-circle route between Los Angeles and Tokyo (34°N and 35°N, respectively) curves far north and skirts the Aleutian Islands.

NOTE: to avoid repetition, I will use the abbreviation PT for the Pythagorean theorem.

ca. 1800 BCE   Two clay tablets found in Mesopotamia, YBC 7289 and Plimpton 322, prove that the Babylonians knew PT.

ca. 600 BCE   In a group of writings known collectively as *sulbas-turas*, an author by the name Baudhayana states, "The rope which is stretched across the diagonal of a square produces an area double the size of the original square." This is the special case of PT for the 45-45-90–degree triangle. A later *sulbastura* by Katyayana states the general case; it also gives instructions how to build a trapezoidal altar, using several Pythagorean triples.

ca. 570 BCE   Pythagoras is born on the island of Samos. After traveling through the ancient world, he founds the Pythagorean school. He discovers that the laws of musical harmony depend on ratios of integers and concludes that *Number Rules the Universe*—the Pythagorean motto.

ca. 540 BCE   The Pythagoreans prove that $\sqrt{2}$ is irrational, and are said to have proved PT. Both proofs are lost.

326 BCE   Alexander the Great conquers the ancient world and founds a new city, Alexandria, in lower Egypt. Its university and famed library attract scholars from all corners of the world.

ca. 300 BCE   Euclid compiles the *Elements*, a summary of the state of mathematics as known in his time. It is to be the most influential mathematical text of all time. PT appears as Proposition I 47, and again as Proposition VI 31 with a different proof. The converse of PT appears as Proposition I 48.

ca. 250 BCE	Archimedes of Syracuse applies PT to a series of inscribed and circumscribing polygons to approximate the value of π; he shows that $3\frac{10}{71} < \pi < 3\frac{10}{70}$.
Second century BCE	Claudius Ptolemaeus, commonly known as Ptolemy, proves *Ptolemy's theorem*: in any cyclic quadrilateral $ABCD$, $AB \times CD + BC \times DA = AC \times BD$. PT follows as a special case when $ABCD$ is a rectangle. The theorem appears in his great work, *Almagest*.
100 BCE–100 CE	Heron of Alexandria proves the formula $A = \sqrt{s(s-a)(s-b)(s-c)}$ for the area of a triangle in terms of its sides and its semiperimeter $s = (a+b+c)/2$ (the formula may have already been proved by Archimedes). Heron's proof was based on proportions; modern books base it on a double application of PT.
The Han Dynasty (206 BCE–221 CE)	PT is known in China as the *kou-ku* theorem. One of the earliest Chinese works on mathematics, the *Chao Pei Suan Ching* (The Arithmetical Classic of the Gnomon and the Circular Paths of Heaven; the exact date is unknown) states PT in words and gives a proof by dissection.
Third century CE	Pappus of Alexandria proves an extended version of PT, true for any triangle.
	Proclus writes his *Eudemian Summary*, a work that includes his own commentary on Book I of the *Elements* and a historical outline of Greek geometry up until Euclid's time.
389	First burning of the library of Alexandria.
ca. 390	Theon of Alexandria writes a revised version of Euclid's *Elements*, from which most modern editions are derived.
641	Second burning of the library of Alexandria.
Ninth century	Tabit ibn Qorra ibn Mervan, Abu-Hasan, al-Harrani proves a generalization of PT involving any triangle.
Eleventh century	Bhaskara gives a "proof without words" of PT identical to the Chinese proof, adding a single word, "see!"

Gherardo of Cremona translates Euclid's *Elements* and Ptolemy's *Almagest* from the Arabic into Latin, making these works accessible to European scholars.

Eleventh to thirteenth centuries	The first European universities are founded: Bologna, in 1088; Paris, 1200; Oxford, 1214; Padua, 1222; and Cambridge, 1231.

1453 Constantinople falls to the Turks, who change its name to Istanbul. This date is considered the end of the Middle Ages.

1454 Johann Gutenberg invents the movable-type printing press.

1482 The first printed edition of the *Elements* appears in Venice.

1570 The first English edition of the *Elements* is published.

1593 François Viète uses a variation of Archimedes' method and discovers the infinite product

$$\frac{2}{\pi} = \sqrt{\frac{1}{2}} \times \sqrt{\frac{1}{2} + \frac{1}{2}\sqrt{\frac{1}{2}}} \times \sqrt{\frac{1}{2} + \frac{1}{2}\sqrt{\frac{1}{2} + \frac{1}{2}\sqrt{\frac{1}{2}}}} \times \cdots .$$

1637 René Descartes invents coordinate (analytic) geometry, in effect unifying geometry and algebra. One consequence is the distance formula between two points in the plane, $d = \sqrt{(x_2 - x_1)^2 + (y_2 - y_1)^2}$.

Pierre de Fermat conjectures that the equation $x^n + y^n = z^n$ has no solutions in positive integers except when $n = 1, 2$. The conjecture became known as *Fermat's Last Theorem* (FLT). The word "last" refers to the fact that FLT was the last of Fermat's conjectures to remain unrproved; it was finally proved in 1994.

1645 Evangelista Torricelli, using the infinitesimal version of PT, rectifies the logarithmic spiral. He shows that the arc length from any given point to the spiral's pole is finite. This is the first known rectification of a transcendental (nonalgebraic) curve.

1649 Euclid's proof of PT features on Laurent de la Hire's painting, *Allegory of Geometry*.

1658	Christopher Wren rectifies the cycloid. He show that the arc length of one arch is equal to eight times the radius of the generating circle.
1666–1676	Isaac Newton and Gottfried Wilhelm Leibniz independently discover the differential and integral calculus. This enabled mathematicians to rectify numerous algebraic and transcendental curves.
1731	First appearance in print of the distance formula in a work on space curves by Alexis Claude Clairaut.
1734	Leonhard Euler proves that the infinite series $1 + 1/2^2 + 1/3^2 + \ldots$ converges to $\pi^2/6$. The proof indirectly involves PT.
1753	Euler proves FLT for the case $n = 3$.
1820	Carl Friedrich Gauss is said to have proposed cutting a huge right triangle, with squares on its sides, out of the forests of Siberia, as a means of signaling to the inhabitants of the Moon our knowledge of PT. Although uncorroborated, the story has surfaced from time to time and was hinted at in Jules Verne's classic science fiction work, *From the Earth to the Moon* (1865).
1828	Julius Plücker introduces line coordinates into geometry. The equation of the unit circle is $\alpha^2 + \beta^2 = 1$, where α and β are the line coordinates of the tangent lines to the circle.
1854	Bernhard Riemann delivers his doctoral address, "On the Hypotheses Which Lie at the Foundation of Geometry," in which he introduces the ideas of multidimensional spaces and curved spaces. He also introduces the *metric* $ds^2 = \sum_{ij} a_{ij} dx_i dx_j$ as a generalization of PT. According to Riemann, every space has its own metric, which can vary from point to point. In effect, Riemann said, the properties of a space are local, not global: each point has its own form of PT.
1876	James A. Garfield, the future twentieth president of the United States, proposes an original proof of PT; the demonstration "hit upon the General in a mathematical discussion with other M.C.'s [Members of Congress] about 1876."

1888	E. A. Coolidge, a blind girl, offers a dissection proof of PT similar to Bhaskara's proof.
1905	Albert Einstein publishes his special theory of relativity, in which the Lorentz transformation plays a central role. The footprints of PT appear in nearly every formula of relativity.
1907	Erhard Schmidt and Maurice Fréchet, expanding on earlier work by David Hilbert, introduce functional spaces, infinitely many dimensional vector spaces in which each "vector" is a function. The "distance" of $f(x)$ from the origin is given by $\sqrt{\int_a^b [f(x)]^2\, dx}$, provided the integral exists. These *Hilbert spaces* would play a crucial role in modern physics.
1908	Hermann Minkowski gives special relativity a four-dimensional interpretation in which the expression $x^2 + y^2 + z^2 + (ict)^2$, where $i = \sqrt{-1}$, c is the speed of light, and t denotes time, is invariant under the Lorentz transformation. He also unifies space and time into one entity, *spacetime*. The distance between two events in spacetime is $\sqrt{(x_2 - x_1)^2 + (y_2 - y_1)^2 + (z_2 - z_1)^2 + (m_2 - m_1)^2}$, where $m = ict$.
1916	Einstein publishes his general theory of relativity, in which Riemann's four-dimensional metric $ds^2 = \Sigma_{ij}\, g_{ij}\, dx_i\, dx_j$ plays a key role.
1927	Elisha Scott Loomis publishes *The Pythagorean Proposition,* which he wrote in 1907 and revised in 1940, the year of his death. The revised edition contains 371 proofs, a "Pythagorean Curiosity," and five Pythagorean magic squares.
1934	Stanley Jashemski, age nineteen, of Youngstown, Ohio, proposes possibly the shortest known proof of PT.
1938	Ann Condit, a sixteen-year-old student at Central Junior-Senior High School in South Bend, Indiana, devises an original proof of PT.
1955	The town of Tigani on the island of Samos (Greece) is renamed Pythagorio. A statue of Pythagoras is built at the town's harbor.

1958	"The Square of the Hypotenuse," a song by Saul Chaplin with lyrics by Johnny Mercer, features in the film *Merry Andrew*.
1993	Andrew Wiles at Princeton University announces he had proved Fermat's Last Theorem. The news appears on the front page of the *New York Times*.
1994	Wiles fixes a flaw found in his 200-page proof. FLT is now considered proven.
1996	Alexander Bogomolny founds a Web site, *The Pythagorean Theorem and Its Many Proofs* (www.cut-the-knot.org/pythagoras/index.shtml). Several more have appeared since.

Aaboe, Asger. *Episodes from the Early History of Mathematics*. New York: Random House, 1964.

Abbott, Edwin A. *The Annotated Flatland: A Romance of Many Dimensions*, with introduction and notes by Ian Stewart. Cambridge, Mass.: Perseus Publishing, 2002.

Ball, W. W. Rouse. *A Short Account of the History of Mathematics*. 1908. Rpt. New York: Dover, 1960.

Baron, Margaret E. *The Origins of the Infinitesimal Calculus*. 1969. Rpt. New York: Dover, 1987.

Boyer, Carl B. *History of Analytic Geometry: Its Development from the Pyramids to the Heroic Age*. 1956. Rpt. Princeton Junction, N.J.: Scholar's Bookshelf, 1988.

———. *A History of Mathematics*. 1968. Rev. ed. New York: John Wiley, 1989.

———. *The History of the Calculus and Its Conceptual Development*. New York: Dover, 1959.

Burton, David M. *The History of Mathematics: An Introduction*. Boston: Allyn and Bacon, 1985.

Cajori, Florian. *A History of Mathematics*. 1893. 2d ed. New York: Macmillan, 1919.

———. *A History of Mathematical Notations*. 2 vols. 1929. Rpt. Chicago: Open Court, 1952.

Coolidge, Julian Lowell. *A History of Geometrical Methods*. 1940. Rpt. New York: Dover, 1963.

Courant, Richard. *Differential and Integral Calculus*. 2 vols. 1934. Rpt. London and Glasgow: Blackie and Son, 1956.

Courant, Richard, and Herbert Robbins. *What Is Mathematics?* 1941. Rpt. New York and Oxford: Oxford University Press, 1996.

Dantzig, Tobias. *The Bequest of the Greeks*. New York: Charles Scribner's Sons, 1955.

———. *Number: The Language of Science*. Ed. Joseph Mazur; foreword by Barry Mazur. New York: Pi Press, 2005.

Devlin, Keith. *Mathematics: The Science of Patterns*. New York: Scientific American Library, 1997.

———. *Mathematics: The New Golden Age*. New York: Columbia University Press, 1999.

Dunham, William. *Journey through Genius: The Great Theorems of Mathematics*. New York: John Wiley and Sons, 1990.

———. *The Mathematical Universe: An Alphabetical Journey through the Great Proofs, Problems, and Personalities*. New York: John Wiley and Sons, 1994.

Euclid: The Elements, translated with introduction and commentary by Sir Thomas Little Heath. 3 vols. New York: Dover, 1956.

Eves, Howard. *An Introduction to the History of Mathematics*. Fort Worth, Tex.: Saunders, 1992.

———. *Great Moments in Mathematics (Before 1650)*. Washington, D.C.: Mathematical Association of America, 1983, lecture 4.

———. *A Survey of Geometry: Revised Edition*. Boston: Allyn and Bacon, 1972.

Frederickson, Greg N., *Dissections: Plane & Fancy*. Cambridge, U.K.: Cambridge University Press, 1997.

———. *Hinged Dissections: Swinging and Twisting*. Cambridge, U.K.: Cambridge University Press, 2002.

Friedricks, Kurt Otto. *From Pythagoras to Einstein*. Washington, D.C.: Mathematical Association of America, 1965.

Gillispie, Charles Coulston, ed. *Dictionary of Scientific Biography*. 16 vols. New York: Charles Scribner's Sons, 1970–1980.

Greenberg, Marvin Jay. *Euclidean and Non-Euclidean Geometries: Development and History*. San Francisco: W. H. Freeman, 1973.

Guggenheimer, Heinrich W. *Plane Geometry and Its Groups*. San Francisco: Holden-Day, 1967.

Heath, Sir Thomas Little. *A Manual of Greek Mathematics*. Oxford: Oxford University Press, 1931.

———. *The Works of Archimedes*. 1897; with Supplement, 1912. Rpt. New York: Dover, 1953.

Heilbron, J. L. *Geometry Civilized: History, Culture, and Technique*. Oxford: Oxford University Press, 1998.

Henderson, Linda Dalrymple. *The Fourth Dimension and Non-Euclidean Geometry in Modern Art*. Princeton, N.J.: Princeton University Press, 1983.

James, Jamie. *The Music of the Spheres: Music, Science, and the Natural Order of the Universe*. New York: Copernicus, 1993.

Jeans, Sir James. *The Growth of Physical Sciences*. Cambridge, U.K., and New York: Macmillan, 1948.

Joseph, George Gheverghese. *The Crest of the Peacock: Non-European Roots of Mathematics*. 1991. Rpt. Princeton, N.J.: Princeton University Press, 2000.

Kaku, Michio. *Hyperspace: A Scientific Odyssey through Parallel Universes, Time Warps, and the 10th Dimension*. New York: Anchor Books, 1994.

Kline, Morris. *Mathematical Thought from Ancient to Modern Times*. 3 vols. New York and Oxford: Oxford University Press, 1990.

Kramer, Edna E. *The Nature and Growth of Modern Mathematics*. 1970. Rpt. Princeton, N.J.: Princeton University Press, 1981.

Krauss, Lawrence M. *Hiding in the Mirror: The Mysterious Allure of Extra Dimensions, from Plato to String Theory and Beyond*. New York: Viking, 2005.

Loomis, Elisha Scott. *The Pythagorean Proposition*. Reston, Va.: National Council of Teachers of Mathematics, 1968.

Mankiewicz, Richard. *The Story of Mathematics*. Princeton, N.J.: Princeton University Press, 2001.

Maor, Eli. *To Infinity and Beyond: A Cultural History of the Infinite*. Princeton, N.J.: Princeton University Press, 1991.

———. *e: The Story of a Number*. Princeton, N.J.: Princeton University Press, 1994.

———. *Trigonometric Delights*. Princeton, N.J.: Princeton University Press, 1998.

Nelson, Roger B. *Proofs without Words: Exercises in Visual Thinking*. Washington, D.C.: Mathematical Association of America, 1993.

———. *Proofs without Words II: More Exercises in Visual Thinking*. Washington, D.C.: Mathematical Association of America, 2000.

Neugebauer, Otto. *The Exact Sciences in Antiquity*. 1957. Rpt. New York: Dover, 1969.

Penrose, Roger. *The Road to Reality: A Complete Guide to the Laws of the Universe*. New York: Alfred A. Knopf, 2005.

Simmons, George F. *Calculus with Analytic Geometry*. New York: McGraw-Hill, 1985.

Singh, Simon. *Fermat's Enigma: The Epic Quest to Solve the World's Greatest Mathematical Problem*. New York: Walker, 1997.

Smith, David Eugene. *History of Mathematics*. Vol. 1: *General Survey of the History of Elementary Mathematics*. Vol. 2: *Special Topics of Elementary Mathematics*. 1923–1925. Rpt. New York: Dover, 1958.

Swetz, Frank J., ed. *From Five Fingers to Infinity: A Journey through the History of Mathematics*. Chicago and LaSalle, Ill.: Open Court, 1995.

Swetz, Frank J., and Kao, T. I. *Was Pythagoras Chinese? An Examination of Right Triangle Theory in Ancient China*. University Park, Penn.: Pennsylvania State University Press, and Reston, Va.: National Council of Teachers of Mathematics, 1977.

Taylor, C. A. *The Physics of Musical Sounds*. London: English Universities Press, 1965.

Toubis, Michael S. A., publisher. *Samos, Icaria & Fournoi: History-Art-Folklore-Routes*. Nisiza Karela, Koropi, Attiki (Greece), 1995.

van der Waerden, Bartel Leendert. *Science Awakening: Egyptian, Babylonian and Greek Mathematics*. 1954. Trans. Arnold Dresden, 1961. Rpt. New York: John Wiley, 1963.

Weisstein, Eric W. *CRC Concise Encyclopedia of Mathematics*. Boca Raton, Fla.: CRC Press, 1999.

Wheeler, John Archibald. *A Journey into Gravity and Spacetime*. New York: Scientific American Library, 1990.

Wilson, Robin J. *Stamping through Mathematics*. New York: Springer-Verlag, 2001.

Yates, Robert C. *Curves and Their Properties*. 1952. Rpt. Reston, Va.: National Council of Teachers of Mathematics, 1974.

Web Sites

Pythagorean Theorem and Its Many Proofs: http://www.cut-the-knot.org/pythagoras/index.shtml

Pythagoras's Theorem: http://www.sunsite.ubc.ca/DigitalMathArchive/Euclid/java/html/pythagoras.html

Ask Dr. Math, High School Archive: http://mathforum.org/library/drmath/drmath.high.html

Pythagorean Theorem—From MathWorld: http://mathworld.wolfram.com/PythagoreanTheorem.html

The Theorem of Pythagoras: http://www.math.uwaterloo.ca/navigation/ideas/grains/pythagoras.shtml

A Proof of the Pythagorean Theorem by Liu Hui: www.staff.hum.ku.dk/dbwagner/Pythagoras/Pythagoras.html

History Topics Index: http://www-gap.dcs.st-and.ac.uk/~history/indexes/HistoryTopics.html

Illustrations Credits

The author is the source for all visual material not credited in the captions or listed here; numbers in parentheses refer to plate and figure numbers.

Color plates

Robin A. Wilson, *Stamping through Mathematics*, courtesy of Springer Verlag, New York (Plate 1)

Richard Mankiewicz, *The Story of Mathematics*, Princeton University Press, Princeton, NJ (Plate 2)

Halftones and Figures:

Asger Aaboe, *Episodes from the Early History of Mathematics*, courtesy of Random House, New York (1.1a, 1.1b)

Anthony Ashton, *Harmonograph: A Visual Guide to the Mathematics of Music*, courtesy of Walker and Company, New York (2.2, 2.10)

Frank Drake and Dava Sobel, *Is Anyone Out There? The Scientific Search for Extraterrestrial Intelligence*, courtesy of Delacorte Press, New York (15.1)

Howard Eves, *An Introduction to the History of Mathematics*, courtesy of Saunders College Publishing, Philadelphia (2.1)

Gilbert W. Grosvenor, ed., *National Geographic Political Globe: Index and Guide*, courtesy of the National Geographic Society, Washington, DC (12.2)

Arthur Koestler, *The Sleepwalkers*, courtesy of Macmillan Co., New York (2.11)

Elisha Loomis, *The Pythagorean Proposition*, courtesy of the National Council of Teachers of Mathematics, Ann Arbor, MI (8.1, 8.2, 8.3, 8.4)

Y. Ladijanski, *Geometry, Part I: Plane Geometry*, n.p., Jerusalem, Israel (3.4)

Richard Mankiewicz, *The Story of Mathematics*, courtesy of Princeton University Press, Princeton, NJ (5.5, 5.9)

Eli Maor, *e: The Story of a Number*, courtesy of Princeton University Press, Princeton, NJ (7.7, 13.1, App.F.1)

Eli Maor, *Trigonometric Delights*, courtesy of Princeton University Press, Princeton, NJ (7.8, 8.5, 8.6, SB 9.2, SB 9.3)

Eli Maor, *To Infinity and Beyond: A Cultural History of the Infinite*, courtesy of Princeton University Press, Princeton, NJ (2.3)

Wolfgang Amadeus Mozart, Piano Concerto No.16 in D Major, K. 451 (in full score), courtesy of Dover Publications, New York (16.1)

Joseph Needham, *Science and Civilisation in China*, courtesy of Cambridge University Press, Cambridge, UK (5.4)

O. Neugebaur, *The Exact Sciences in Antiquity*, courtesy of Dover Publications, New York (1.3)

New York Cares, promotional image, newyorkcares.org (S8.1)

Dale Seymour et al., *Line Designs*, courtesy of Creative Publications, Palo Alto, CA (10.9)

David Eugene Smith, *History of Mathematics, Vol. 1: General Survey of the History of Mathematics*, courtesy of Ginn and Company, Boston (2.2, 5.11, 5.12)

Spiegel, *Mathematical Handbook of Formulas and Tables*, courtesy of McGraw-Hill Book Company, New York (16.2)

TWA, promotional image, n.p. (S9.1)

Note: Arabic names with the prefix *al* are listed alphabetically according to their main name, preceded by *al-*; for example, al-Biruni is to be found under the letter B.

The main text in this book is set in Times Roman, a serif typeface commissioned by *The Times of London* newspaper in 1931. It was designed by Stanley Morison (1889–1967) to address the problems of high-speed printing on low quality newsprint. Although no longer used by *The Times*, it is still widely used in book typography.

The headers and display text are set in Bodoni, a serif typeface designed by Giambattista Bodoni (1740–1813) in 1798. Many variants have been designed since. Bodoni forms the basis of a number of corporate identities, among them CBS and IBM print advertisements.

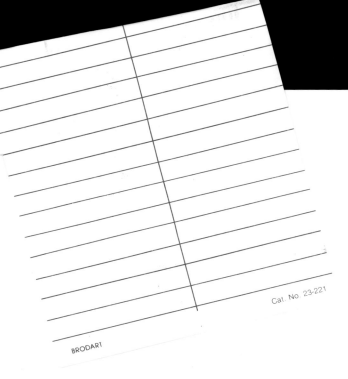